Rogers Locomotive and Machine Works, Matthias Nace Forney

Locomotives and Locomotive Building

Rogers Locomotive and Machine Works, Matthias Nace Forney

Locomotives and Locomotive Building

ISBN/EAN: 9783744692007

Printed in Europe, USA, Canada, Australia, Japan

Cover: Foto ©berggeist007 / pixelio.de

More available books at **www.hansebooks.com**

Thomas Rogers

LOCOMOTIVES

—AND—

LOCOMOTIVE BUILDING,

BEING

A BRIEF SKETCH OF THE GROWTH OF THE RAILROAD SYSTEM AND OF THE
VARIOUS IMPROVEMENTS IN LOCOMOTIVE BUILDING IN AMERICA
TOGETHER WITH A HISTORY OF THE

ORIGIN AND GROWTH

OF THE

Rogers Locomotive and Machine Works,

PATERSON, NEW JERSEY,

FROM 1831 TO 1886.

J. S. ROGERS, *Pres't.*
R. S. HUGHES, *Sec'y.* } PATERSON, N. J.
~~JOHN HEADDON, *Sup't.*~~

R. S. HUGHES, *Treas'r.*
44 Exchange Place,
NEW YORK.

NEW YORK:

WM. S. GOTTSBERGER, PRINTER, 11 MURRAY STREET,

1886.

PREFACE.

The last catalogue of the Rogers Locomotive and Machine Works with a sketch of the origin and growth of that establishment, was published in 1876.

Since then many changes have been made in the equipment of these Works and in the character, design, and dimensions of the locomotives turned out. To describe these adequately it was necessary to rewrite nearly the whole of the former volume. This work was entrusted to my hands by the officers of the Rogers Locomotive and Machine Works. As it was commenced during the fiftieth year that the establishment had been engaged in the manufacture of locomotives, it seemed a suitable time to give a somewhat full account of the origin and history of the Works, and of the evolution of the locomotives built in them during that period. Such an account has been carefully prepared, and consists very largely of what may be called a mechanical history of the work which has been done; which, it is thought, will be interesting to many readers, as it shows the successive steps which have led to the wonderful development of the locomotive in this country. It also indicates the extent to which the perfection of the modern American type of locomotive is due to the ingenuity, mechanical skill, and sound judgment of the founder of this establishment — Mr. Thomas Rogers, and to his successor — Mr. William S. Hudson. Both of them have left a record of their genius and ability in their designs, which are imitated to-day, and which promise to survive until locomotives are superceded.

Very complete data concerning the dimensions and performance of the locomotives which this establishment is now prepared to furnish are given by illustrations and tables in the latter part of the book, and as there is still considerable difference of opinion and practice in calculating the capacity of locomotives, an explanatory chapter is given showing just how the calculations were made.

M. N. FORNEY.

NEW YORK, October 1, 1886.

CONTENTS.

THE ROGERS
LOCOMOTIVE AND MACHINE WORKS.

CHAPTER I.

THE ROGERS LOCOMOTIVE AND MACHINE WORKS were founded by Thomas Rogers, who was born March 16th, 1792, in the town of Groton in New London County, Connecticut. He died in New York City, April 19th, 1856. He served in the war of 1812, and was a lineal descendant of Thomas Rogers, one of the Pilgrim Fathers, who came over to this country from England in the Mayflower. At the age of sixteen he was apprenticed to learn the trade of a house carpenter, and in the summer of 1812 he removed to Paterson, N. J., then a small village which at that time was very prosperous on account of the demand for American manufactures brought about by the war with Great Britain. Many of the manufacturers were reduced to bankruptcy on conclusion of peace, in 1815.

At this time he was employed as a journeyman carpenter, and was noted for his constant application to business, good judgment, and force of character. A few years afterward, Captain Ward, who had been travelling in Europe, where he had seen the power-loom in operation, came to Paterson for the purpose of introducing the manufacture of cotton duck. Mr. Rogers was employed to make the patterns for these looms. He very soon understood their construction and recognized their value and bought from Captain Ward the patent right for making them.

In 1819, he associated himself with John Clark, Jr., under the firm name of Clark & Rogers. They commenced work in the basement story of the Beaver Mill, a building which at an early day had been put up by Mr. Clark's father. Shortly afterwards, Mr. Rogers visited Mexico, where he received large orders for looms, etc. In 1820 the firm moved into the little Beaver Mill, and in the following year took into partnership Abraham Godwin, Jr., and the firm name was then changed to Godwin, Rogers & Co. They then commenced spinning cotton and building machinery for that and other purposes.

In 1822, finding their accommodations too limited, they leased Collett's Mill and moved into it. Their business continued to increase, the number of persons employed being sometimes as high as 200. The establishment continued to prosper until the summer of 1831. In the latter part of June of that year Mr. Rogers withdrew, and took with him $38,000 as his share of the profits of the firm.

1

He then took a mill-site on the upper raceway in Paterson, and immediately commenced the erection of the " Jefferson Works," which were finished and put in operation before the close of the following year. The location and building of the " Jefferson Works " was literally an encroachment on the forest. On the upper race no factories had been put up, except two little cotton mills and a small machine shop, the latter owned by Messrs. Paul & Beggs. Between Spruce and Mill streets, all was swamp covered with pines.

It was the intention of Mr. Rogers to devote the lower stories of the " Jefferson Works " to building machinery, and the upper stories to spinning cotton. The latter was, however, never commenced, as the demand for machinery increased so fast that the whole of the new building was devoted to that branch of the business.

In the early part of 1832, he associated with himself Messrs. Morris Ketchum and Jasper Grosvenor, of New York, the name of the firm being Rogers, Ketchum & Grosvenor.

In that year the railroad from Jersey City to Paterson was approaching completion, and the iron work for the bridges over the Passaic and Hackensack rivers had been made by Mr. Rogers. An order was also executed for one hundred sets of wheels and axles for the South Carolina railroad, of which Mr. Horatio Allen was then chief engineer. A short time before Mr. Allen had visited England to get information about the use of locomotives on railroads, and at the time he ordered the work for the South Carolina Railroad he recommended Mr. Rogers to undertake the construction of locomotives.

In the following letter, written more than fifty years after the event, Mr. Allen describes his interview with Mr. Rogers:

SOUTH ORANGE, N. J., Dec. 31, 1884.

Dear Sir : —

" The earliest railroad work in this country was done by the West Point Foundry Association to which was entrusted the order for railroad wheels for the South Carolina Company, and other work for that Company.

" Knowing that the Era that had opened would require works specially appropriate to the construction of the rolling stock up to the locomotives, I obtained authority in the spring of 1830 from the South Carolina Railroad Company to seek the works which in position, instrumentalities, and preparedness, were in condition to undertake and were willing to undertake what was wanted.

" The result of inquiries to the end in view led me to call on Rogers, Ketchum & Grosvenor, a firm then engaged in the manufacture of machinery for cotton and woolen mills, whose works were at Paterson, N. J.

" At these works I called and asked an interview with Mr. Rogers, the partner having charge of all the mechanical operations of the firm. It was without any letter of introduction or any personal knowledge of each other. My subject was my introduction, and Mr. Rogers very soon led me to know that I had come to the right place and to the right man.

" At the close of an hour's conversation Mr. Rogers expressed his readiness to enter the new field, and to undertake any orders that were entrusted to their firm. The future of ' The Rogers Locomotive Works ' was determined at that hour's conversation.

" The personal and business relations which followed this interview, continued for many years, and were to me of the most satisfactory character."

Yours truly,

HORATIO ALLEN.

Rogers Locomotive Works

One of the accompanying engravings represent the works of Rogers, Ketchum & Grosvenor, as they were in 1832, and the other shows them as they are in 1886.

The following advertisement, which first appeared in the *American Railroad Journal* of June 8, 1833, will give an idea of the character of the business of the firm at that time

RAILROAD CAR WHEELS AND BOXES,
AND OTHER RAILROAD CASTINGS.
☞ Also, AXLES furnished and fitted to wheels complete, at the Jefferson Cotton and Wool Machine Factory and Foundry. Paterson, N. J. All orders addressed to the subscribers at Paterson, or 69 Wall street, New-York, will be promptly attended to, Also, CAR SPRINGS.
Js ROGERS, KETCHUM & GROSVENOR.

This advertisement was continued regularly until December 24, 1836.

CHAPTER II.

THE EARLY HISTORY OF RAILROADS IN THIS COUNTRY.

IN 1833 railroads were already attracting a great deal of attention in this country. The opening of the Erie Canal for commercial purposes in 1826, and the consequent diversion of traffic from other seaboard cities to New York, led the people of Philadelphia, Baltimore, Boston and Charleston to seek for means by which their lost trade could be recovered. Investigation and accurate surveys soon showed the impracticability of constructing canals from Baltimore to the Ohio River, or from Boston to the Hudson. In the meanwhile information concerning the successful use of steam power on the Stockton & Darlington Railroad in England, which was opened in 1825, had reached this country, and the public had received the reports of the celebrated experiments with locomotives which were made on the Liverpool & Manchester Railway in 1829. As Mr. Charles Francis Adams, Jr. has expressed it :— *

"America suffered from too few roads ; England from too much traffic. Both were restlessly casting about for some form of relief. Accordingly all through the time during which Stephenson was fighting the battle of the Locomotive, America, as if in anticipation of his victory, was building railroads. . . .

"The country, therefore, was not only ripe to accept the results of the Rainhill contest, but it was anticipating them with eager hope."

After the experiments referred to had been made, full reports giving in detail their results, were published in this country, Committees of inquiry were sent to England

* See Railroads : their Origin and Problems.

to get information and report on the railroads of that country, and a railroad mania began to pervade the land.

The first railroad which was built in the United States was a short line of about three miles from the Quincy granite quarries to the Neponset river, * for the transportation of granite for the Bunker Hill Monument. This was merely a tram road and was operated by horse power and stationary engines, and was built in 1826. As Mr. Adams says :—

"Properly speaking, however, this was never — or at least, never until the year 1871, — a railroad at all. It was nothing but a specimen of what had been almost from time immemorial in common use in England, under the name of 'tramways.'"

A similar work was constructed at about the same time for the transportation of coal from the pits mouth to the Lehigh Valley Canal near Mauch Chunk, Pa.

In the latter part of 1827 the Delaware & Hudson Canal Company put the Carbondale railroad under construction. This road extends from the head of the Delaware and Hudson Canal at Honesdale, Pa., to the coal mines belonging to the Delaware & Hudson Canal Company at Carbondale, a distance of about sixteen miles. This line was opened, probably, in 1829, and was operated partly by stationary engines, and partly by horses. The line is noted chiefly for being the one on which a locomotive was first used in this country. This was the Stourbridge Lion (Fig. 2,) which was built in England under the direction of Mr. Horatio Allen, who had been an assistant engineer on this line. It was tried at Honesdale, Pa., in August 1829.

Fig. 2.
"STOURBRIDGE LION," 1829.

According to Poor's Railroad Manual for 1876 and 1877 : " It was not until 1828, that the construction of a railroad was undertaken for the transportation both of freight and passengers on anything like a comprehensive scale. The construction of the Erie Canal had cut off the trade which Philadelphia and Baltimore had hitherto received from the West ; and as the project of a canal from the city of Baltimore to the Ohio was regarded by many as impracticable, the merchants of that city, in 1827, procured the charter of the present Baltimore & Ohio Railroad. On the 4th of July, 1828, the construction of the railroad was begun, the first act being performed by the venerable Charles Carroll, of Carrollton, the only then surviving signer of the Declaration of Independence. At the close of the ceremony of breaking ground, Mr. Carroll said :—

" I consider this among the most important acts of my life, second only to that of signing the Declaration of Independence, if even second to that."

* It has recently been stated that as early as 1809 an experimental railroad track, 180 feet in length, was laid in Delaware County, Pa., and that in the same year a road about a mile long was constructed from stone quarries on Crum Creek to a "landing" on Ridley Creek in the same county and state. The evidence upon which this statement is based has not been made public.

" In the fall of 1829, the laying of the rails within the City of Baltimore was begun. On the 22d of May, 1830, the first section of fifteen miles, to Ellicott's Mills, was opened.

" The next important railroad was the South Carolina,* begun in 1830, and opened for traffic in 1833 for its whole length (135 miles). At that time, it was the longest continuous line of railroad in the world. The construction of the Mohawk & Hudson Railroad, now a part of the New York Central, was begun in 1830. It was opened (17 miles) in 1831. The Saratoga & Schenectady Railroad (21½ miles), was opened in the following year; the Paterson & Hudson River Railroad was chartered in January, 1831, construction on it was commenced in 1832, and it was opened in 1834; the Cayuga & Susquehanna (34 miles), connecting the " Susquehanna River with the Cayuga Lake, was opened in 1834; and the Rensselaer & Saratoga (25 miles) in 1835. In New Jersey, that portion of the Camden & Amboy, extending from Bordentown to Hightstown, (14 miles) was opened on the 22d of December, 1830; and between Hightstown and South Amboy (47½ miles) in 1834. In Pennsylvania a considerable extent of line for the transportation of coal had been constructed previous to 1835. In 1834 the Philadelphia & Columbia (82 miles) and the Portage Railroad (36 miles), both forming a part of the system of public works undertaken by the State of Pennsylvania, were opened. The completion of these gave that State a continuous line, made up of canal and railroad, from Philadelphia to the Ohio River at Pittsburgh. The total mileage of railroad constructed in the State of New York up to, and including, 1835, was 265 miles, or more than one-quarter of the whole extent of line then in use in the United States. In 1833 the Baltimore & Ohio Railroad was extended as far west as Harper's Ferry (81 miles). In the same year the Washington branch (30 miles) was also completed. In Massachusetts, in 1835, the Boston & Worcester Railroad (44 miles); the Boston & Providence (41 miles), and the Boston & Lowell (26 miles) were all opened for business. The total mileage in operation in all the States at the close of that year was 1,098 miles."

The preceding sketch of the early history of railroads, in this country, is given to show the extent of railroad construction at the time that Mr. Rogers determined to undertake the manufacture of locomotives.

---

* The original charter of the South Carolina Railroad was granted Dec. 19, 1827. This was not satisfactory to some of the citizens of Charleston, and a new bill was reported to the legislature on the 22d of January 1828 and passed on the 29th of the same month. The stockholders organized as a company on the 12th of May, 1828.

CHAPTER III.

THE EARLY HISTORY OF LOCOMOTIVES IN THIS COUNTRY.

IN the latter part of the year 1827, the Delaware & Hudson Canal Company decided to have built in England three locomotives, for their line of railroad from Honesdale to Carbondale. This action was taken on the report of the Chief Engineer of the road, Mr. John B. Jervis, and Mr. Horatio Allen, who had been an engineer on the line, went to England and was authorized to have the engines built on plans to be decided by him while there. He arrived in England in 1828, and ordered one engine from Foster Rastrick & Co., of Stourbridge. This was the Stourbridge Lion, (Fig. 2.). Two other engines were ordered from Stephenson & Co., of Newcastle.

In a pamphlet with the title "The Railroad Era," written by Mr. Allen in 1884, he says:—

"The two locomotives from Stephenson that were in New York early in the year 1829, and therefore prior to the trial of the locomotive "Rocket" in October of that year, were identical in boiler, engines, plan and appurtenances with the "Rocket" (Fig. 3.); and if one of the two engines in hand ready to be sent had been the one used on August 9th 1829, the performance of the "Rocket" in England would have been anticipated in this country."

"The three locomotives were received in New York in the winter of 1828 and 1829. One of each kind was set up, with the wheels *not* in contact with the ground, and steam being raised, every operation of the locomotive was fully presented except that of onward motion."

Fig. 3.

None of these engines were sent to the road for which they were intended, until the following spring. The Stourbridge Lion, so far as is known, was the only one which was ever placed on the road. It was not tried until August 9th 1829, and was then run by Horatio Allen, who has the honor of being the first person who ever ran a locomotive in America.

This engine, it was said, was too heavy for the road, and was used only a short time. It is a singular fact that it is not now (1886) known what became of the two engines, built by Stephenson & Co., and which were in every essential similar to the celebrated "Rocket."

In August 1830, Peter Cooper tried his "model of experimental locomotive engine," (represented by Fig. 4.) on the Baltimore & Ohio Railroad. This engine had but one working cylinder of $3\frac{1}{4}$ in. diameter, and $14\frac{1}{2}$ in. stroke of piston. The engine was tried on August 28th, 1830. In the same year the South Carolina Railroad Company contracted with Mr. E. L. Miller, to build a locomotive, which was named the Best Friend, for the South Carolina Railroad Company. This engine, (shown by

Fig. 4.   Fig. 5.

Fig. 5.), was put into service in November 1830, and was the first locomotive ever built in America for actual service upon a railroad.

A locomotive called "The South Carolina," (Fig. 6.), designed by Horatio Allen, was built for the South Carolina Railroad by the West Point Foundry Association, in the year 1831. The boiler had its fire-box in the middle, with a pair of barrels (four in all) extending each way, with a chimney at each end. The engine had eight wheels,

Fig. 6.

arranged in two trucks, one pair of driving wheels, and one pair of leading wheels forming a truck. Each truck had one cylinder which was in the middle of the engine and attached to the smoke-box. The driving axle had a crank in the middle to which the connecting rod was attached by a ball-joint. The trucks were connected to the engine by king-bolts in the usual way.

The "De Witt Clinton," (Fig. 7.) was the third locomotive built by the West Point Foundry Association. It was made for the Mohawk & Hudson Railroad, and was ordered by John B. Jervis, Esq. The first excursion trip with passengers, drawn by the De Witt Clinton, was made from Albany to Schenectady, August 9th, 1831.

On January 4th 1831, the Baltimore & Ohio Railroad offered the sum of $4,000 "for the most approved Engine which shall be delivered for trial upon the road on or before the 1st of June 1831 — and $3,500 for the Engine which shall be adjudged the next best."

Fig. 7.   Fig. 8.

Three or four locomotives, amongst them one with a rotary engine, built by Mr. Childs of Philadelphia, entered into the competition during the summer of 1831. The only one of them, named the "York," which proved equal to the moderate performance required of them, was the one built by Messrs. Davis & Gartner, two machinists of York, Pa. The engine had a vertical boiler and vertical cylinder; with four coupled wheels 30 inches in diameter. It was altered considerably after being placed on the road. The Atlantic was afterwards built by the same firm, and was the first of what were afterwards known as the grasshopper engines, (Fig. 8,) which were used for many years on the Baltimore & Ohio Railroad.

Fig. 9.

In August 1831, the locomotive, John Bull, (Fig. 9.) built by George & Robert Stephenson & Co., of Newcastle upon Tyne, was received in Philadelphia for the Camden & Amboy Railroad & Transportation Company. This is the old engine which was exhibited at the Centennial Exhibition in Philadelphia in 1876. In the winter of 1831 or 1832, three locomotives built by the same firm in England were received and were put to work on the Newcastle & Frenchtown Railroad in Delaware.

The third edition of Wood's Treatise on Railways, published in 1838, contains a tabular statement which gives the names and dimensions of engines built by R. Stephenson & Co., Newcastle upon Tyne, and the names of the railways for which they were built. This table contains the names of the following locomotives for American roads :—

Delaware,   for Newcastle & Frenchtown Railroad.
Maryland,   "   "   "   "   "
Pennsylvania,   "   "   "   "   "
No. 42, for Saratoga & Schenectady Railroad.
H. and Mohawk, for Mohawk & Hudson Railroad.
Stevens, for New York.
No. 52, for United States.
Edgefield, for Charleston & Columbia Railroad.
Brother Jonathan, for Mohawk & Hudson Railroad.
No. 61,   "   "   "   "   "
No. 75,   for Saratoga & Schenectady Railroad.
Wm. Aikin,   "   Charleston & Columbia   "
No. 99.   "   "   "   "   "
No. 104,   "   Pennsylvania   "
No. 105,   "   "   "   "
No. 106,   "   Columbia   "

No dates are given in the table, but all of these sixteen engines must have been built before 1838. Most of them were probably of what was known as the " Planet " class shown by Fig. 10., which is the form of engine, that succeeded the " Rocket," and the only one which the Stephensons built for some years after its adoption. These locomotives which were imported from England, doubtless, to a very considerable extent, furnished the types and patterns from which the engines which were afterwards built here, were fashioned. But American designs very soon began to depart from their British prototypes and a process of adaptation to the existing

Fig. 10.

conditions of the railroads in this country followed, which afterwards " differentiated " the American locomotives more and more from those built in Great Britain. Until recently a marked feature of difference between American and English locomotives has been the use of the truck, under the former. Its use was proposed by Mr. Horatio Allen, in a report dated May 16, 1831, which he made to the South Carolina Canal & Railroad Company, of which he was then the chief engineer. The locomotive with two trucks, shown by Fig. 6, was built from his design in the latter part of 1831, and was put into operation on the South Carolina Railroad in the early part of 1832. In the latter part of the year 1831 the late John B. Jervis invented what he called "a new plan of frame, with a bearing carriage, for a locomotive engine, for the use of the Mohawk & Hudson Railroad, represented by Fig. 11, which was constructed and put on the road in the season of 1832."

A truck was also devised by Ross Winans and applied to a locomotive on the Baltimore & Susquehanna Railroad (now the Northern Central) in the latter part of 1832. In a letter published in the *American Railroad Journal* of July 27, 1833, Mr. Jervis describes the objects aimed at in the use of the truck as follows:—

"The leading objects I had in view, in the general arrangement of the plan of the engine, did not contemplate any improvement in the power over those heretofore constructed by Stephenson & Co.;* but to make an engine that would be better adapted to railroads of less strength than are common in England; that would travel with more ease to itself and to the rail on curved roads; that would be less effected by inequalities of the rail, than is attained by the arrangement in the most approved engines."

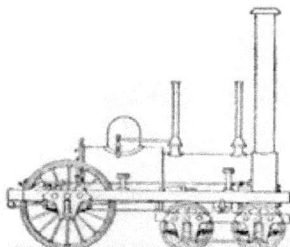

Fig. 11.

The effectiveness of the truck in accomplishing what it was intended for was at once recognized, and its almost general adoption on American locomotives followed.

In the year 1833, Judge Dickerson, then President of the Paterson & Hudson River Railroad, ordered a locomotive, which was called the "McNeill," from George Stephenson, which was to be as good as possible without regard to cost. It arrived, and was put in operation in the year 1835. The cylinders were 9 inches diameter by 18 inches stroke, and the engine had one pair of driving wheels five feet in diameter, which were behind the fire-box. The axle was cranked, and the cranks were close to the wheels; there was room for the connecting rods to pass by the outside of the furnace. The front end was supported by a four-wheeled truck; the fire-box and tubes were of copper. The engine continued in use many years, and was said to be very fast and was finally sold to a western railroad, the business of the Paterson & Hudson River Railroad, having grown beyond the engine's capacity.

There may have been other English engines, of which there is no record, imported into this country about this time, but, as already stated, there is no doubt that to a very considerable extent the English engines were the models from which American designers received many suggestions; but, as will be shown, they very soon began to depart from the original types, and the development of the locomotive here was quite distinct from that which it had in Europe.

* The truck was applied by Mr. Jervis to an engine built by Stephenson & Co., of England.

CHAPTER IV.

HISTORY OF LOCOMOTIVE BUILDING AT THE ROGERS LOCOMO-
TIVE AND MACHINE WORKS.

PREPARATION for locomotive building in Paterson had been made as early as 1833 by Messrs. Paul & Beggs, in their shop near that of Mr. Rogers. They had a small engine nearly completed when their building took fire and was consumed, and the locomotive destroyed.

In 1835 some buildings were begun by Messrs. Rogers, Ketchum & Grosvenor, with a view to the manufacture of locomotives. The following notice and advertise-ment, which appeared in the *American Railroad Journal* of Dec. 24, 1836, will give an idea of the character of the business of the firm at that time :

AMERICAN LOCOMOTIVES.

" By the following advertisement we learn — and it affords us pleasure to call to it the attention of our readers interested in railroads — that Messrs. Rogers, Ketchum & Grosvenor, of Paterson, New Jersey, have added to their extensive machine shops one for Locomotive Engines.

" We have more than once enjoyed the pleasure of a visit to their works, where we found ample evidence of the truth of a remark often made by us, that 'to whatever branch of manufacture our countrymen turn their attention they are sure to excel,' and so, we doubt not, it will be in this new branch of business undertaken by this enterprising house, and we hope soon to learn that their skill in this branch has been as successful as in others.

" In a few years we shall not see an imported Locomotive on an American Railroad."

The following is the advertisement referred to :

MACHINE WORKS OF ROGERS,

KETCHUM AND GROSVENOR, Paterson, New-Jersey. The undersigned receive orders for the fol-lowing articles, manufactured by them, of the most superior description in every particular. Their works being extensive, and the number of hands employed being large, they are enabled to execute both large and small orders with promptness and despatch.

RAILROAD WORK.

Locomotive Steam-Engines and Tenders ; Driv-ing and other Locomotive Wheels, Axles, Springs and Flange Tires ; Car Wheels of cast Iron, from a va-riety of patterns, and Chills ; Car Wheels of cast iron, with wrought Tires ; Axles of best American refined Iron ; Springs ; Boxes and Bolts for Cars.

COTTON WOOL AND FLAX MACHINERY.

Of all descriptions and of the most improved Pat-terns, Style and Workmanship.

Mill Gearing and Millwright work generally ; Hy-draulic and other Presses ; Press Screws ; Callen-ders ; Lathes and Tools of all kinds, Iron and Brass Castings of all descriptions.

ROGERS, KETCHUM & GROSVENOR
Paterson, New-Jersey, or 60 Wall street, N. Y.
5tf

The first locomotive, the Sandusky, Fig. 12, which the firm built, was not completed until 1837. It was intended for the New Jersey Railroad & Transportation Company. The engine was 4 ft. 10 in. gauge, the same as that of the line for which it was built. It had cylinders 11 in. diameter by 16 in. stroke, with one pair of driving wheels of 4 ft. 6 in. diameter, which were placed in front of the fire-box. The engine had a truck in front with four 30 in. wheels. The cylinders were inside the frames and were connected to a crank axle of the form shown in Fig. 13. The eccentrics were outside of the frame, and the eccentric rods extended back to rocking shafts which were located under the footboard. The smoke pipe was of the bonnet kind, and had a deflecting

Fig. 12.

cone in its centre. The edges of the cone were curled over so as to deflect the sparks downward, and thus prevent their passing through the wire bonnet, as well as preventing the bonnets from wearing out too fast.

The driving wheels of the engine were made of cast iron, with hollow spokes and rim, which at the time was a remarkable novelty. The section of the spokes was of an oval form and the rim of very much the same shape as that which is in common use at the present time. This kind of driving wheel has since come into almost universal use in this country.

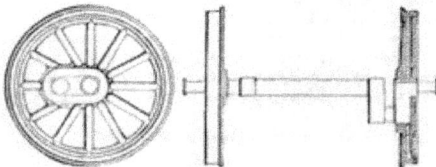

Fig. 13.

Another important improvement adopted by Mr. Rogers in the construction of this engine, was the counterbalancing the weight of the crank, connecting rods and piston. For this he filed a specification in the Patent Office, dated July 12, 1837. It is described as follows in the specification:

"The nature of my improvement consists in providing the section of the wheel opposite to the crank with sufficient weight to counterbalance the crank and connecting-rods, making the resistance of the engine less in starting, and in running; also, preventing the irregularity of motion caused by that side of the wheels when the cranks are placed in the usual mode of fitting them up. The irregular motion which arises from not having the cranks and connecting-rods balanced, is attended with much injury to the engine, and to the road, and with much loss of power."

In order to counterbalance the weight of the parts referred to, the rim of the wheel opposite the crank was cast solid, while the other part of it was made hollow. The importance of counterbalancing was not recognized until several years after it had been introduced by Mr. Rogers, and, when attention was drawn to it, many doubted the necessity of balancing anything more than the cranks.

The trial trip of the Sandusky was made from Paterson to Jersey City and New Brunswick and back on the 6th of October, 1837, Mr. Timothy Smith acting as engineer. The performance of the engine was entirely satisfactory; the gauge of the road was 4 ft. 10 in., the same as that of the New Jersey Railroad & Transportation Company, for which road the engine was intended. It was, however, bought for the Mad River & Lake Erie Railroad by its President, Mr. J. H. James, of Urbana, Ohio, and on the 14th, it was shipped via Canal and Lake, in charge of Mr. Thomas Hogg, in the schooner "Sandusky." Mr. Hogg had worked upon it from the commencement. It arrived at Sandusky, Nov. 17, 1837, at which time not a foot of track had been laid. The road was built to suit the gauge of the engine, and the Legislature of Ohio passed an Act requiring all roads built in that State to be of 4 ft. 10 in. gauge, the same as the engine Sandusky.

The engine was used in the construction of the road until the 11th of April, 1838, when regular trips for the conveyance of passengers commenced between Belleview and Sandusky, a distance of 16 miles.

The engineer was Thomas Hogg, who ran the engine for three years, keeping it in repair. It continued in service many years, until engines of larger size were required to do the work.

The second locomotive built by Mr. Rogers was called the "Arresseoh No. 2." It was completed in February 1838 for the New Jersey Railroad & Transportation Company. It was similar in design to the "Sandusky."

The third engine was named the "Clinton" and was built for the Lockport & Niagara Falls Railroad Company, and was delivered to it in April 1838. It differed from the first engines in having cylinders which were 10 in. in diameter and 18 in. stroke and the gauge was 4 ft. 8½ in. Both the driving and the truck wheels of this engine had hollow oval spokes and hollow rims with wrought iron tires. This engine was run by Wm. E. Cooper until November 1843, when it was sold to the Toledo & Adrian Railroad for $6,500, the original cost. It was said by Mr. Cooper that when the engine was sold it was considered to be one of the best working engines in existence.

An engine called the "Experiment," was the next, or the fourth locomotive turned out. It was made for the South Carolina Railroad, and was delivered in June 1838. This engine differed from those previously built at these works, in having a smaller cylinder and longer stroke than usual.

The Sandusky was the type of the first four locomotives built by Messrs. Rogers, Ketchum & Grosvenor. In many respects they all resembled the Stephenson engines. They had inside cylinders and a crank-axle but differed from English locomotives chiefly in having a truck instead of a pair of leading wheels. The driving axles were in front of the fire-boxes, with the result that the overhang of the latter behind the axle brought an undue proportion of the weight of the engine on these axles.

To remedy the evil of an excessive amount of weight on the driving axle the latter was placed behind the fire-box in the fifth engine, called the "Batavia," Fig. 14, built at these works. When this was done, however, there was too little load on the driving wheels, and an arrangement was provided for transferring part of the weight of the tender to them. The Batavia was built for the Tonawanda Railroad, and was completed in 1838.

The shape of the furnace, in plan, was semi-circular at the rear part, and it had a hemispherical top surmounted with a dome. This form of fire-box was used as late as 1857.

In his early engines, besides using inside cylinders Mr. Rogers also followed the plan which is still used in England, viz: putting the cranks for parallel or coupling rods

Fig. 14.

opposite to the main cranks. He soon found that this arrangement, while it had some advantages, such as requiring less counterbalance, caused the journals of the driving axles to wear oval; he therefore adopted the plan of putting the cranks for both main and outside rods on the same side of the centre of the axle.

The "state of the art" of locomotive building in this country in its infancy is graphically described in the following articles, which appeared in the American Railroad Journal and Mechanic's Magazine of Dec. 15, 1839. In one of these the editor said:

"A few days ago, in company with one of the proprietors, we had the pleasure of a visit to, and inspection of the very extensive works of Messrs. Rogers, Ketchum & Grosvenor, at Paterson, New Jersey, for the construction of various kinds of machinery. Our attention was, of course, principally directed to the shops for the construction of locomotives, the main building of which is 200 feet long and three stories high, and another of equal length containing near 50 forges, most of which were in operation, notwithstanding the pressure of the times.

"We saw a number of engines in different states of forwardness, and though the general forms are those of 6-wheeled American Engines in general, we were not a little gratified with several minor arrangements, new to us at least, which have been introduced by Mr. Rogers, and to which we shall briefly refer.

"The wire gauze of the smoke pipe is protected by an inverted cone, placed in the axis of the pipe, a few inches below the wire gauze. The base of the cone is curled over so as to scatter the sparks over a large portion of the surface of the wire cloth, and to prevent the top of the spark-catcher from being burnt out before the rest of the wire cloth is materially injured; it also tends to throw the larger sparks down between the pipe and the casing, and will do something towards diminishing this standing reproach.

"The truck frames, whether of wood or iron, were admirably stiffened by diagonal braces, and where the crank axle is used, the large frame is very strongly plated in the manner of Stephenson's engines, the neglect of which till very lately has been, we are informed, a constant objection to the Philadelphia engines on the Long Island and Troy railroads.

"The wheels are of cast iron, with wrought iron tires; the spokes are round, and they, as well as the rims, are hollow, except where the crank axle is used, when the rims are cast solid on one side so as to counterbalance the cranks.

"Our readers will probably remember an article on this subject in the *Journal*, Nos. 7 and 8, page 244 of the present volume, on "side motion or rocking," by G. Heaton, where its success on the Birmingham railroad has been complete.

"Mr. Rogers balanced his first engine wheels two and a half years since, and entered a specification, not with the intention of taking out a patent, but to prevent anyone else from doing so; and thus deprive the community of the benefit which Mr. Rogers was desirous of conferring, and which we understand other makers are now availing themselves of. The advantages are fully explained in the article referred to.

"When the crank axle is used, the eccentric rods and the cranks of the rockshafts are placed on the outside, where they are easily got at, and where they are not crowded into the smallest possible space, as with the ordinary arrangement. For this, also, a specification was entered with the same object as in the preceding case.

"But we were most pleased with the arrangement of levers to which the eccentric rods are fastened, and thus the reversing depends on no contingency, for the rods are forced in and out of gear; a single handle only is required to manage the engine much more rapidly and efficiently than by the ordinary mode. The boilers are 8 ft. long for an 8-ton engine, and with 120 flues, the usual length of the former being, we believe, 7 ft., and the number of the latter about 80 or 90; by this deviation the area of heating surface is increased, and the heat remains longer in contact with the flues, while the addition to the weight is very trifling compared with the advantages derived from the saving of fuel.

"Mr. Baldwin, of Philadelphia, took out a patent some time since for a very ingenious mode of saving half the crank, by inserting the wrists into one of the spokes of the driving wheels, and this has been very closely imitated by making one complete crank, and by letting one-half of it into a spoke which is cast larger than the others, with a receptacle for the purpose. This latter plan has been adopted by Mr. Rogers and others in this neighborhood, whilst the Boston machinists aim at bringing the two cranks as near together as possible. The relative merits of straight and cranked axles are so well pointed out in Mr. Wood's papers on locomotives in these numbers, that we shall merely beg leave to state that the plan of Mr. Baldwin and its imitation, appear to us to combine the liability to fracture of the crank axle with the loss of heat, the exposure to accident, and the racking of frame and road ascribed to the straight axle; for the only difference is the thickness of the spoke, the loss of heat is the same in both, the protection against any serious accident is too trifling to be considered, whilst, with the cranks as close together as possible, the cylinders are completely protected.

"We offer these remarks as our views merely, and with all due deference to the superior skill of Messrs. Baldwin and Rogers. Mr. Rogers, in common with all other experienced machinists with whom we have conversed, is decidedly opposed to any increase of width of track beyond 5 ft., with the present weight of engine.

"As regards the power of the engines, they are able to slip the wheels when the rails are in the best state; this they do in common with all good American or English engines, consequently any accounts of extraordinary performance would be worse than superfluous, when we know that they will do all that any other engine whatever, with the same weight on the driving wheels, possibly can do.

"As a last remark, we would observe, that there is more finish on the engines of Messrs. Rogers, Ketchum & Grosvenor than we are in the habit of seeing; some parts usually painted black being highly polished. On the whole we consider their new establishment eminently calculated to add to the reputation of American locomotives, as it has for many years largely contributed to the character of American machinery for the manufacture of cotton and other objects."

AN EXTRAORDINARY FEAT.

In the same number of the same journal, is the following letter which still further elucidated the subject:

"GENTLEMEN.— As you seem to take a deep interest in the success of American locomotives, I will give you a statement for your gratification, in relation to a performance on the New Jersey Railroad a few days since.

"Owing to some circumstances, of which I am not informed, it became necessary for a locomotive on the way from Jersey City to New Brunswick, to take, in addition to its own load, the cars attached to another engine, which made the number equal to 24 loaded four-wheeled cars, and with as much apparent ease as could be desired, notwithstanding the grade for four miles is equal to 26 ft. per mile, stopping on the grade to take in passengers, and starting again with the greatest ease. The average speed on the grade was 24½ miles per hour.   This may not be in your estimation anything extra-ordinary, yet I consider it a performance worth recording, by way of contrast with the greatest and most extraordinary performance of a locomotive ever heard of in these days, which occurred on the Liverpool & Manchester Railroad in 1829, only ten years ago.   Twenty tons on a level road at the rate of ten miles per hour, was then considered wonderful! Astonishing!   Even in a country famed for its extraordinary discoveries; yet here, only ten years after, we see an engine built in this country too, taking a load probably equal, cars and tender included, to 120 or 180 tons at the rate of 24½ miles per hour, up a grade of 26 ft. per mile.   This engine was built, I understand, at Paterson, New Jersey, by Messrs. Rogers, Ketchum & Grosvenor, a concern not yet so well known to this railroad community as manufacturers of locomotives as they ought to be, or as they soon will be, if they continue to turn out such machines as the one above alluded to.

"If such have been the improvements in the past, what may they not be, permit me to ask, in the next ten years?

"Pardon me for thus troubling you, but my aim is rather to call attention to the rapid march of improvement in this mode of communication, than to direct attention to any individual or company, although those gentlemen, in my opinion, deserve as manufacturers, much more than I have said of them.

"Yours truly,

NEWARK, N. J., December 14, 1839.   "JERSEY BLUE."

Soon after he commenced building locomotives Mr. Rogers became convinced that inside connected engines, with crank axles, were inferior in many respects to outside connected ones, besides being more expensive to build and to keep in repair; he also became satisfied that in the matter of steadiness, the inside-connected had no advantage over the outside-connected engine, and that, with proper counter-balancing, the latter could be run as fast as required without any injurious oscillation; and also, that it required more skill to properly counterbalance inside connected engines than outside ones.   Therefore, he was an earnest advocate of this style of engine, and recommended outside-connected engines as better than inside-connected ones.

Fig. 15 represents the "Stockbridge," built in 1842, with outside cylinders.   In this engine the driving axle was placed in front of the fire-box and a pair of trailing wheels behind to carry the overhanging weight.   The load on the driving wheels was of course reduced by an amount equal to that carried by the trailing wheels, so that this type of engine was also deficient in adhesion and power.

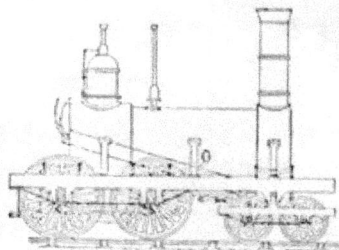

Fig. 15.   Fig. 16.

The next step which was made was to substitute a pair of driving wheels for the trailing wheels, and couple them with the main driving wheels.   This form of engine, shown by Fig. 16, was patented in 1836 by Henry R. Campbell, of Philadelphia, and was adopted by Mr. Rogers in 1844. This plan has since been so generally adopted in this country that it is now known as the "American" type. Fig. 17 represents an engine of this kind built at the Rogers Works in 1844. It had four coupled driving wheels and outside cylinders, the eccentrics were on the back axle, the pumps were full stroke, worked from the cross-heads. It had springs over the back axle bearings, and also in the centre of the levers which extended from the driving axle to the centre of the truck on each side of the engine. The truck was pivoted and turned upon a centre pin fixed to

Fig. 17.   Fig. 18.

the boiler; the arrangement did not give satisfaction, and was altered after a short trial. This engine was remarkable from the fact that it is the first example of the use of an equalizing beam between the driving wheels and truck.

The engine shown by Fig. 18 was built in 1845, and had equalizing levers between the driving wheel springs; the truck had side bearings and springs over the sides of truck; the pumps had short stroke and were worked from the cross-head as shown.

Fig. 19.

Fig. 19 shows an engine built in 1846 with the driving wheels spread well apart. It had V hooks and independent cut-off on the back of the main valves; this was a favorite kind of engine for many years.

In 1848 Mr. Rogers was requested to furnish some engines with six coupled wheels for the Savanilla Railroad in Cuba. He then designed and built the first ten wheeled engines ever made at the Rogers Works. There is no drawing of these engines extant. They had, however, outside cylinders 15½ in. diameter by 20 in. stroke. The ten-wheeled engines, which had been built previous to this time, had inside cylinders and crank axles. The connecting rods of the engines for the Savanilla Railroad were made to take hold of the outside journal of the main crank pin, which at that time was a new departure.

Fig. 20 represents a plan of ten-wheeled engine, with half-crank keyed on the driving wheel, same as Baldwin's plan. This pattern of engine was built in 1848 after those for the Savanilla Railroad. The engine had outside bearings and equalizing levers between the springs; it also had cranks on the axles outside the frames to which the coupling rods were attached. A number of engines on this plan, with cylinders 17 × 22, was built for the New York & Erie Railroad. They all had independent cut-off valves.

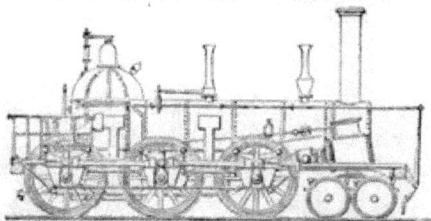

Fig. 20.

Fig. 21 represents an inside cylinder engine with full crank; the steam chests were inclined sidewise, so that the valves could be readily got at. This was one of the improvements introduced by Thomas Rogers. The engine had V hooks and independent cut-off valves, and was built for the Paterson & Hudson River Railroad.

Fig. 21.

On the style of engine shown by Fig. 22, the shifting link motion was introduced. Thomas Rogers was one of its earliest advocates, and did more towards its successful introduction on American locomotives than any other person. He was not only an early, but an earnest advocate of it, at a time when it was condemned by some of the most prominent engineers in the country. Time has amply proved all that he claimed for it, which was that it is the most simple and efficient form of valve gear that has ever been devised.

Fig. 22.

Fig. 23 represents a style of passenger engine which was first built in 1852. It had 15 × 22 in. cylinder driving wheels 5 ft. in diameter. It had what may be called

Fig. 23.

supplementary outside frames, which carried the running board, cab, &c. It had shifting links, hung from below, and the truck axles had both inside and outside bearings.

The form of engine represented by Fig. 24, was first built in 1853, and was for a

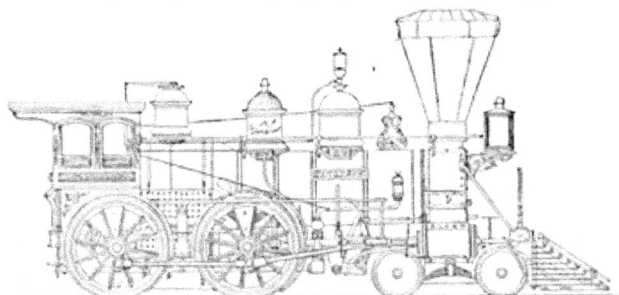

Fig. 24.

long time very popular. Many railroads in the country were equipped with them. The cylinders were 16 × 22 in. and the driving wheels 5 ft. diameter, although the size of the latter was varied somewhat in different engines.

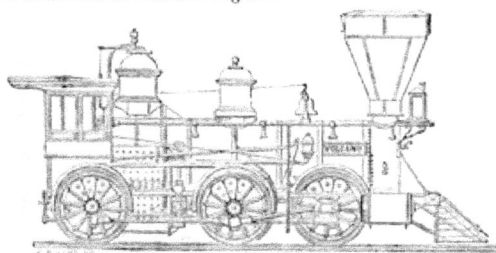

Fig. 25.

Fig. 25 represents a six-wheeled coupled engine built in 1854. The following report of its performance was published in the *American Railway Times* in 1859 :

"The engine 'Vulcan,' of the Buffalo & State Line Railway, came out of the shop after a general overhauling, on the 15th of December, 1856, and made 15 trips of 90 miles each, 1,350 miles, and hauling 435 cars in that month.

"In the year 1857, this engine made 312 trips of 90 miles each, hauling 8,509 cars; in the year 1858, this engine made 290 trips, hauling 9,351 cars."

On the death of Mr. Thomas Rogers, which occurred in 1856, the business theretofore conducted by Rogers, Ketchum & Grosvenor was re-organized under a charter, with the title of The Rogers Locomotive & Machine Works, and Mr. William S. Hudson was then appointed Superintendent. He was a prolific inventor and an excellent mechanic, and introduced many improvements in locomotive construction, which will be described further on.

The first " Mogul " engine, Fig. 26, built at the Rogers Works, was completed in 1863. This plan of locomotive was made possible by the invention of the Bissell truck and the addition of the swing links to it by A. F. Smith, both of which will be described in another chapter. With a single axle truck in front of the cylinder, the front driving

Fig. 26.

wheels can be placed farther forward than they can be on a ten-wheeled engine with a four-wheeled truck, one axle of which is in front, and another behind the cylinders. Consequently Mogul engines have a larger proportion of their weight on the driving wheels than ten-wheeled engines have, and this has brought the Moguls in favor for freight service.*

The demand for more powerful locomotives naturally suggested coupling four pairs of wheels and led to the " consolidation " type, which has eight driving wheels coupled, and a pony truck in front of the cylinders. In 1880 the first consolidation engine built at the Rogers Works was completed, and was substantially like that shown by plate VI.

The types of engines which have been described, are the principal ones which have been evolved in this country for ordinary freight and passenger service. Besides

---

* A plan shown in Plate X was designed for a ten-wheeled engine at the Rogers Locomotive Works with a four-wheeled truck in front of the cylinder. The order for these engines was however, ultimately given to another establishment. In this design it was aimed to secure all the advantages of both the ten-wheeled and Mogul plans.

Wm. S. Hudson

these there has been a demand for locomotives for special service, such as switching, urban and suburban traffic, and for narrow guage railroads ; the narrowness of which made it essential to design special methods of construction.

The most common plan used for switching engines is that shown in Plate XIII, which has four coupled wheels, both axles being placed between the furnace and smoke-box. Separate tenders are furnished with locomotives of this kind, or the tanks may be placed on top of the boilers as shown in Plate XV.

When more powerful engines are required, six coupled wheels are used with the axles all between the furnace and smoke-box, as in Fig. 25 and Plates XIV and XVI. Some six coupled engines have been built with an axle behind the fire-box, but with this arrangement the overhanging weight of cylinder, smoke-box, &c., bring an undue amount of weight on the front pair of wheels.

The advantage of locating the driving axles between the furnace and smoke-box, is that the overhanging weight of the furnace behind, balances that of the cylinders, smoke-box, &c., in front, and in this way the driving wheels carry the whole weight of the engine and it is equally distributed on them. Placing the water tank on top of the boiler is inconvenient and unsightly, and when in that position it is difficult to get room enough for an adequate supply of water, and there is also the disadvantage of a varying load on the driving wheels, which may be excessive with the tank full, and insufficient when it is empty. For these reasons Mr. Hudson, after he became Superintendent of the Rogers Works, turned his attention to devising methods of construction which would retain all the advantages of the arrangement of axles described, but which would at the same time give a longer wheel base for steadiness, but with sufficient flexibility to enable the engine to run round sharp curves easily. The requirements of suburban and other traffic, in which engines must make short runs, had also created a demand for locomotives which could be conveniently and safely run both ways, and which would not require to be turned around at the end of each journey. Having these objects in view, Mr. Hudson, in 1867, designed and patented the plan of tank locomotive represented by Plate XVII, which soon became known as Hudson's " Double Ender." In this the two driving axles were placed between the furnace and smoke-box, and a Bissell truck was placed at each end of the engine. Mr. Hudson's patent was dated May 7, 1867, and was re-issued December 7, 1875.

It will be seen that the water tank of these engines was on top of the boiler. This arrangement was open to the objections which have been pointed out. To overcome these Mr. Hudson, in 1872, designed and patented the plan of engine represented by Plate XVIII. In this the arrangement of the driving axles and the front truck, excepting the equalizing arrangements, are the same as that of the " Double Ender " plan, but instead a two-wheeled Bissell truck behind, a four-wheeled swing motion truck was substituted, and the water tank instead of being placed on top of the boiler, was placed over the four-wheeled truck. This arrangement was patented July 16, 1872.

In 1866, Mr. M. N. Forney patented the plan embodied in the engine shown in

Plate XX. A number of engines of that kind have been built at the Rogers Locomotive Works for various roads. Whether a leading truck is essential for engines of this class has been a subject of a good deal of controversy among railroad engineers. To reconcile the views of the various parties to this dispute, the Rogers Works build locomotives either with or without the leading truck, as required, leaving to the purchaser and user the task of determining whether a leading truck is useful or not.

In 1872 Mr. Hudson took out seven patents for different plans of tank engines with trucks at each end. In all of them his system of equalizing levers between the trucks and driving wheels springs, which is described in another chapter, was used, and his patents were chiefly for various applications of that system.

Plate XXIII represents an engine built in accordance with one of his patents. It was built for a narrow guage road, and in order to get as wide a fire-box as possible the frames were made as shown by Figs. 180 and 181 and described on page 62.

He also patented in 1873 a plan for a compound locomotive. This had two outside cylinders in the usual position, the one being of larger diameter than the other. It was intended that ordinarily live steam from the boiler should be admitted to the small cylinder only, from which it exhausted into a super-heater in the smoke-box before it passed into the large cylinder on the opposite side. The steam pipe was connected with the steam chest of the large cylinder by another pipe of smaller diameter. Live steam could be admitted by the small pipe to the large cylinder if required. This plan was never put into practice.

Mr. Hudson's death occurred on the 20th of July, 1881. He was then 72 years old.

The following extracts are taken from an account of his life, which appeared in the *Railroad Gazette* immediately after his death:

" He was born near the town of Derby, England, in 1809, and at an early age began to learn the trade of an engineer and machinist, serving part of his apprenticeship under George Stephenson. In 1833, when 24 years of age, he came to this country, and for a time found work in the engine room and machine shops attached to the Auburn State Prison in New York. He soon left that place, however, and engaged as a locomotive runner on the old Rochester & Auburn Railroad, now a portion of the New York Central. Subsequently he ran an engine on the Attica & Buffalo Railroad, and was made Master Mechanic of the road, which he left in 1852 to become Superintendent of the Locomotive Works of Rogers, Ketchum & Grosvenor, at Paterson, N. J. In 1856 these works were incorporated as the Rogers Locomotive and Machine Works, and Mr. Hudson was made Mechanical Engineer and Superintendent, a position which he held until his death. He succeeded Mr. Thomas Rogers, who was the founder of these works, and who probably did more than any other man to develop the design and improve the construction of the American Locomotive as it is to-day. But Mr. Hudson took up the work where Mr. Rogers left it, and during the 30 years that Mr. Hudson occupied the position of the head of the mechanical department of this establishment, he made many improvements in the locomotives built there, chiefly of a kind which are the result of simplifying details, adopting better methods of putting work together, and making the engines more substantial and more serviceable. He studied, as probably no other locomotive builder did the performance of the engines he built. He was constantly looking out for their weak points, and it was said by the present head of the establishment that Mr. Hudson was always more concerned about building a good engine than he was in making a good profit."

The business of the Rogers Locomotive and Machine Works is now conducted by Mr. J. S. Rogers, the President of the Company, who is a son of the founder of the establishment.

CHAPTER V.

THE ORGANIC DEVELOPMENT OF THE LOCOMOTIVE.

DURING the period of fifty years that has elapsed since Mr. Rogers first commenced to build locomotives in Paterson, not only has the machine as a whole been going through a process of evolution, as described in preceding chapters, but there has also been a development or adaptation of its various parts or organs, as they may be called, to the functions which they have to perform. A description of the different forms and methods of construction of these organs, which were adopted and in use at various times, will therefore become a sort of comparative anatomy of American locomotives. This may conveniently be divided into three parts,— one relating to the boiler, another to the engines, and a third to the carriage or running gear. These will be taken up in succession.

THE BOILER.

The boiler of the Sandusky, the first engine built by Messrs. Rogers, Ketchum & Grosvenor, was substantially the same as that of the Stephenson engines, of what is known as the " Planet " class, that is the top of the furnace was semi-cylindrical in form and flush or nearly flush, with the top of the barrel of the boiler. The horizontal section of the fire-box below the barrel of the boiler was square or nearly so.

In 1837 Mr. Bury was made locomotive Superintendent of the London & Birmingham Railway in England, which gave him an opportunity of adopting extensively on that line a class of engines, the original of which he introduced on the Liverpool & Manchester Railway in 1830. These were four-wheeled engines with inside cylinders, not unlike Stephenson's in their general plan, but the tops of the furnaces instead of being semi-cylindrical were hemispherical, and the horizontal section of the fire box, below the waist of boiler, was of a form approximating to the letter D. the flat part being in front. This form of fire-box was adopted in the fifth engine built at the Rogers Works, and it was in continuous use until 1857, and is shown in Figs. 14 to 22.

A large proportion of the early locomotives built in this country were built to burn wood. The Baltimore & Ohio Railroad was perhaps the only pioneer road that commenced by using coal for fuel, and even on that line many locomotives burned wood. As the weight of locomotives was increased and coal was substituted for wood, larger fire-boxes were required, and this led to the abandonment of the hemispherical topped furnace, which was not well adapted to fire-boxes whose length was materially greater than their width, and the semi-cylindrical form which was first used, was substituted in its place. In these the crown sheets were usually stayed with crown-bars placed either lengthwise or crosswise on top of the fire-box.

Fig. 27.   Fig. 28.

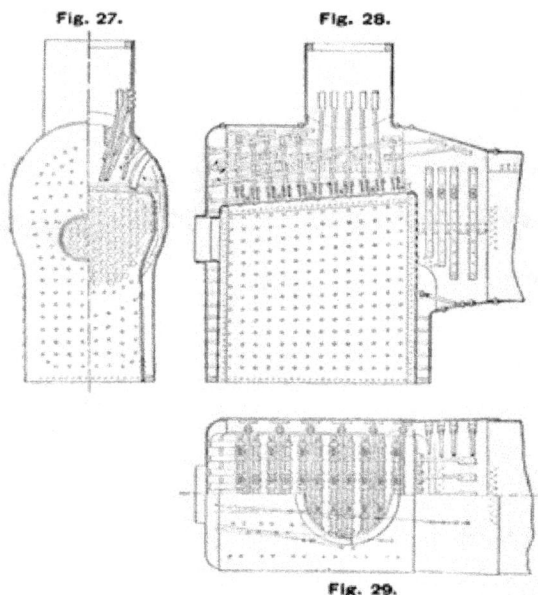

Fig. 29.

At first the cylindrical tops of the furnaces were made flush with the tops of the barrels of the boilers, but this form was succeeded by what is known as the "wagon top" form of boiler, which was first used in the Rogers Works in 1850. The tops of the furnaces, in boilers of this kind, were also semi-cylindrical, but they were made considerably higher than the barrels of the boilers as shown in Figs. 23 to 26. The exact reason for first adopting this form of boiler is not known, but it had the advantage of giving more steam room, and allowed the use of more tubes and consequently more heating surface than could be used in a flush topped boiler. The wagon top also gives more room for workmen on the inside of the boiler, over the crown sheets, and it thus facilitates construction and repairs. Mr. Hudson was always a strong advocate of this form, and he gave especial attention to staying it, as is shown in Figs. 27, 28, and 29, in which the stays and braces are shown.

For burning anthracite coal, it was found that very long fire boxes were required. In 1860 the form shown in Figs. 30 and 31 was built at the Rogers Works from the

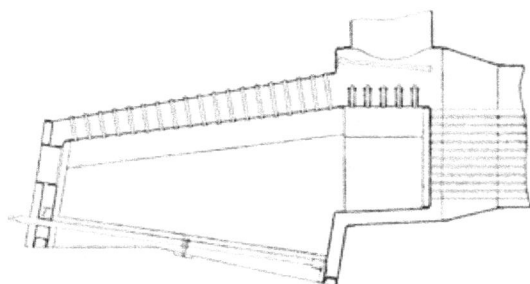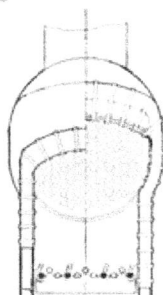

Fig. 30.　　　　　　　　　　　　　　Fig. 31.

design of Mr. Millholland, of the Philadelphia & Reading Railroad. The top of this

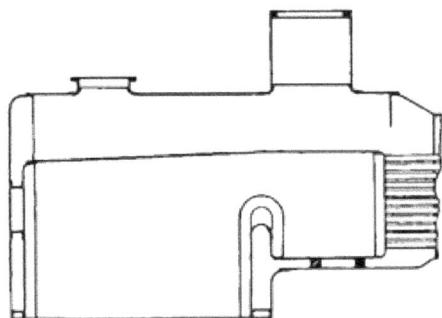

Fig. 32.

furnace sloped downward from the barrel of the boiler, and the crown sheet was stayed with screw stays, excepting for a short distance behind the tube plate. Water grates were used in this fire-box and are shown in the engraving.

In 1861 some fire-boxes with long combustion chambers and a water bridge, as shown in Fig. 32, were constructed for the New Jersey Railroad & Transportation Co.

In 1862 a fire-box with the water leg A, Figs. 33 and 34 was made for the Chicago, Burlington, & Quincy Railroad.

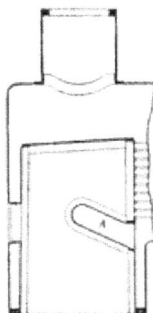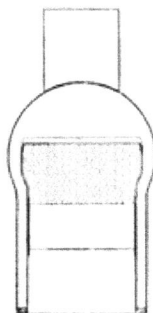

Fig. 33.　　　　　　　Fig. 34.

The brick arch, Figs. 35 and 36, was used in 1865.

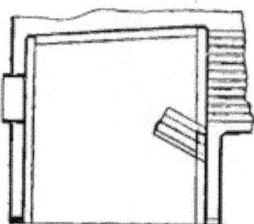

Fig. 35.   Fig. 36.

In 1871 some engines were built for the Cumberland Valley Railroad, with the Buchanan fire-box, shown by Figs. 37 and 38.

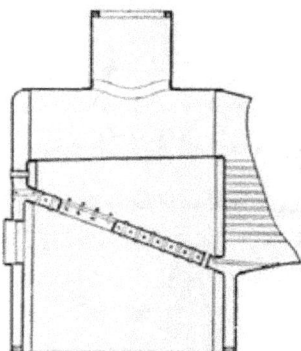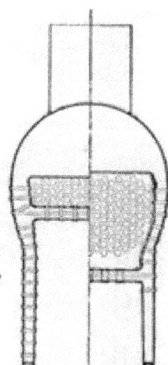

Fig. 37.   Fig. 38.

The form of the Belpaire fire-box, shown by Figs. 39 and 40, was applied to locomotives for the Matanzas Railroad of Cuba in 1874.

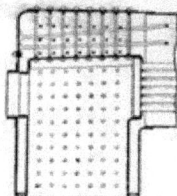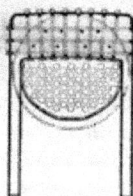

Fig. 39.   Fig. 40.

The Belpaire fire-box has been extensively used on the continent of Europe, and within the past few years has been regarded with much favor by some of the leading master mechanics in this country, and it has been adopted on a number of railroads here.

The fire-box represented by Figs. 27, 28, and 29 is however, the one which has been the most commonly used for engines built at the Rogers Works. It has stood the test of long experience, and is still regarded with much favor by engineers and master mechanics.

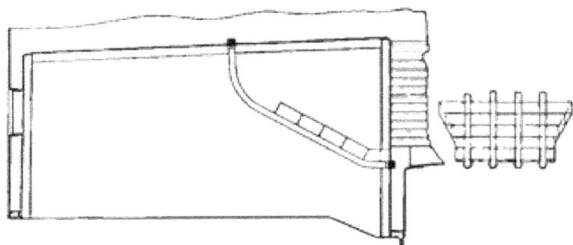

Fig. 41.   Fig. 42.

The form of brick arch shown in Figs. 41 and 42 was used in 1881. In this it will be seen that the fire-brick is supported on bent water tubes which are attached at one end to the crown sheet, and at the other to the front plate of the fire-box. Another form of brick arch supported on water tubes is shown in Figs. 43 and 44. This was used in 1885.

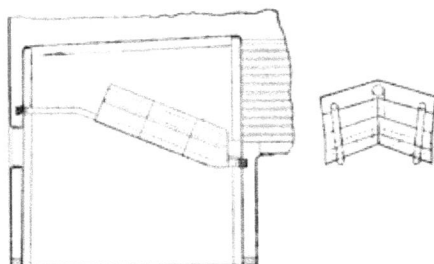

Fig. 43.   Fig. 44.

TUBES.

Very soon after coal was substituted for wood as fuel in locomotives, the use of copper and brass tubes was abandoned in this country, and iron tubes were used instead. At first there was a great deal of trouble in keeping these tubes from leaking. This was especially the case before steam gauges were generally used. Without these instruments it was impossible to tell what the steam pressure was, until the safety valves commenced blowing off. They were therefore the principal guides by which the fireman was governed, that is, he would "fire" until "she commenced blowing off," and then he would open the furnace door wide to cool the fire. The result was that the tubes were thus exposed to alternate currents of cold and hot air, and were thus continually expanded and contracted, which caused them to leak. With a steam gauge, however, a fireman had always a guide before him to indicate just what the steam pressure was, and could control his fire accordingly, and therefore was not obliged to open the furnace door so often to regulate the steam pressure.

While the frequent expansion and contraction of the tubes probably caused them to leak, yet there can be no doubt that the methods of fastening them which were at first used were much less efficient than those which have since been adopted.

Fig. 45.

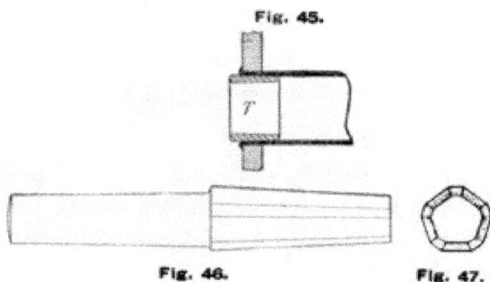

Fig. 46.   Fig. 47.

The manner of fastening tubes in 1837 is shown in Figs. 45, 46, and 47. The tube was inserted into the hole in the tube plate, and a tapered mandril, shown by Fig. 46, was driven into the end of the tube, so as to expand it to the full size of the hole in the plate. This mandril was flattened on five sides, as shown in the end view, Fig. 47. After each blow on the end of the mandril it was turned slightly so as to expand the tube equally all around. The end of the tube was then turned over, as shown in Fig. 45, which represents a longitudinal section of it. Probably some form of caulking tool was used for this purpose. A wrought iron thimble T was then driven into the end of the tube.

Fig. 48.

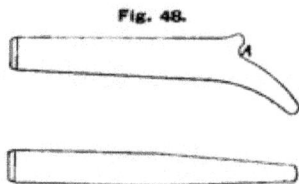

Fig. 49.

In 1840 the form of caulking tool shown in Figs. 48 and 49 was adopted. This was inserted in the end of the tube with the notch A, bearing against the edge, which was then turned over by driving the tool against it with a hammer.

As already stated, thirty or forty years ago a great deal of trouble was experienced on locomotive engines with leaky flues. It was a constant source of annoyance, and every few days some one had to go into the furnace to hammer or caulk up the ends of the flues and thimbles (the flues at that time were either copper or brass, and the thimbles were of wrought iron).

In 1850 Mr. Hudson, then Master Mechanic of the Attica & Buffalo Railroad, conceived the idea that if cast iron thimbles were substituted for wrought iron it would remedy this standing difficulty. Acting on this idea he proceeded to verify it, — first by taking a thimble of each kind, wrought and cast iron, turning them accurately to a guage, then heating them red hot, measuring them, and noting the expansion of each; afterward cooling them in water and again measuring them. This process of heating, cooling, and measuring was repeated twelve times, when the wrought thimble was found to be appreciably smaller in size than at first, and the cast iron thimble larger. It was noticed that the former thimbles expanded more than the latter when red hot; this was anticipated.

To carry this idea into practice, a locomotive with leaky flues was taken :— All the thimbles were taken out, the flues carefully expanded, and new thimbles put in. One half, or all on one side of the centre line of the flue sheet vertically, were of wrought iron, and the other half were all of cast iron. At the end of the first trip, when the boiler was cooling down, it was found that all the flues with wrought iron thimbles were leaking, whereas, at the same time, all those opposite to them with cast iron thimbles were tight. The wrought thimbles were then taken out and cast iron ones put in their places, when all stopped leaking and so continued, the engine doing duty, without any more trouble from leaky flues. The attention of Thomas Rogers was called to the fact, and he began to use cast iron thimbles with a like result. Mr. Rogers called the attention of John Brandt, then in charge of the motive power of the Erie Railway to the subject; he, also, immediately tried cast iron thimbles, and found the result as stated above, and hence their use spread and became almost universal; few, except those who had experience in the matter at that time, can now realize how much annoyance and expense were saved by the change.

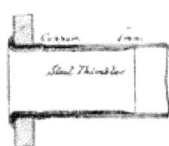

Fig. 50.   Fig. 51.

In 1861 tubes were fastened as shown in Fig. 50, that is, a copper end or thimble was brazed to the end of the tube, and a steel thimble was placed on the inside of it, so as to bring the copper between it and the tube plate. The soft copper between the steel thimble and the plate, it was found, assisted materially in making and keeping the tubes tight.

In 1862 the method shown in Fig. 51 was adopted. In this the copper end was dispensed with and a copper thimble was placed on the end outside of the tube as shown.

The Prosser expander was first used at the Rogers Works in 1863. This is shown by Figs. 52 and 53. Fig. 52 is a side view with the end of the tube and plate shown in section at *A* and *A*. The expander consists of what may be called a plug composed of eight sector-shaped pieces as shown in the end view, Fig. 53. These are held together by an open steel spring ring *B*. In the centre of the sectors there is a tapered hole *C*, Fig. 53 (shown by dotted lines in Fig. 52), into which a tapered plug, Fig. 54, is driven. The open spring ring permits the sectors to separate when the tapered plug is driven into the opening. The sectors each have a shoulder or projection at *S*, *S*. These come just inside the tube plate, when the expander is inserted into the tube. By driving in the tapered plug or mandril, Fig. 54, the sectors are forced apart, and expand the end of the tube. At the same time the shoulders *S*, *S*, produce a ridge in the tube, inside of the plate, which helps to keep the joint tight.

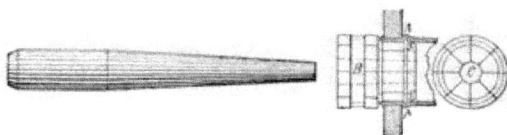

Fig. 54.   Fig. 52. Fig. 53.

In 1867 the Dudgeon expander, shown by Figs. 55 to 58 was introduced. This may be described as a hollow plug which has three rollers, *R*, *R*, *R*, Figs. 55 and 56, which are contained in cavities in the plug in which they can revolve, and in which they can also move a short distance radially, that is, from the centre of the plug outwards.

Fig. 58.   Fig. 57.   Fig. 55.   Fig. 56.

When this expander is inserted in the end of a tube a tapered mandril, Fig. 57, is driven into the central opening, and it then bears against the rollers *R*, *R* and forces them outwards against the tubes. A crank handle is then attached to the square end of the mandril and it is turned around, which causes the rollers to revolve on their own axis. This causes the hollow plug to revolve around its axis. The two thus have a sort of sun and planet motion in relation to each other. As the rollers bear hard against the tube their effect is to elongate it circumferentially, and thus enlarge it so as to completely fill the opening in the tube plate. Usually copper ferrules are used outside of the ends of the tubes. This method is the one which is now generally employed at the Rogers Works for fastening tubes in their plates.

FURNACE DOOR.

In 1865 Mr. Hudson used the furnace door deflector illustrated by Figs. 59 and 60. *D* is the deflector which is suspended from a hook *H*, attached to the fire-box over the furnace door. A lever *L* is fastened to the deflector by which it is moved out of the way when coal is thrown in the fire. The position of the deflector is regulated by the

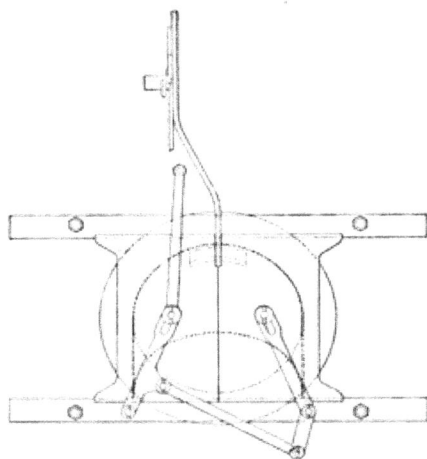 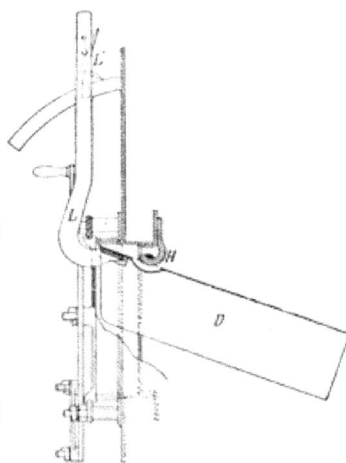

Fig. 59.   Fig. 60.

lever, and a latch *L* at its upper end. A pair of sliding doors are used in connection with the deflector. These are opened by a system of levers which are clearly shown in the engravings. This was first suggested by a fireman in England, who found that by inserting a scoop shovel upside down in the furnace door he could prevent smoke.

BOILER SHELLS.

In making boilers with iron plates, Mr. Hudson always took great pains to have the plates of such sizes and proportions that the "grain" or fibres of the iron around the barrel of the boiler would be in the direction to resist the greatest strain. This practice is still continued in the Rogers Works when iron plates are used.

In 1852 he adopted the method of making the horizontal seams of boilers, shown by Figs. 61 and 62. This consisted of an ordinary single riveted lap seam with a covering strip or " welt " over the inside, which was made wide enough to take an extra row of rivets on each side of the main row. The outside rows were spaced double the distance apart of those in the main row. The welts not only serve to strengthen the seams, but they cover the inside caulking edges where corrosion and " grooving " or " channelling " as it is called, is most likely to occur. By being covered, these edges are protected from the action of the water.

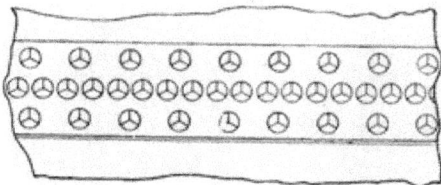

Fig. 61   Fig. 62.

DOMES.

The first method of fastening domes, as shown in the engraving of the Sandusky, Fig. 12, was to rivet a circular casting having a flange, top and bottom, to the barrel of the boiler. The upper part of the dome was also made of cast iron and was bolted to the top flange of the circular casting. A similar plan was also adopted when the domes were attached to tops of the hemispherical shaped furnaces as shown in Figs. 12 to 22.

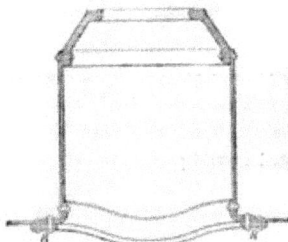

Even after the use of the hemispherical shaped furnace was abandoned, cast iron domes were still used, and in some cases the bases of the domes were made of wrought iron. When the size of engines and their domes was increased so much that it became impracticable and unsafe to make them of cast iron, they were made of wrought iron plates, with a flange at the bottom, which was riveted to the boiler shell as shown in Fig. 30. Later the boiler shell was flanged upward around the edge of the opening at the base of the dome, as shown in Figs. 26 and 27, in order to give additional strength at this point. The dome was then attached to the boiler with two rows of rivets. In 1880 a reinforcing ring R, R, was added at the base of the dome as shown in Fig. 63. This serves to strengthen the boiler shell at the base of the dome, where it is weakened by the opening required to give access to the inside of the boiler.

Fig. 63.

GRATES.

With very few exceptions, the fuel used in the early locomotives in this country was wood. This could be burned successfully with an ordinary " plain " grate as it was called, consisting of narrow bars with spaces about ½ in. wide between them. Figs. 64 and 65 show a grate of this kind, which was used in 1840. The bars were made of cast iron, the material of which locomotive grates are almost universally made in this country. Figs. 66 and 67, however, represent a grate made of wrought-iron bars, bolted together in groups of four bars each. The use of wrought-iron bars is however an exception to the general practice in this country. The grate shown in the figures last referred to has a drop door *D* at the front end. This is hinged at *B* and is held up by the arms *A, A,* on the shaft *S.* To drop the door, the shaft is turned by the lever on the end of the shaft which lowers the arms *A, A,* and allows the door to fall.

Fig. 66.

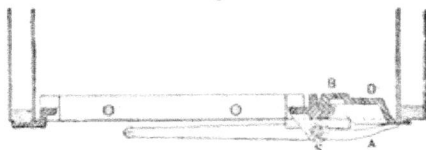

Fig. 64.

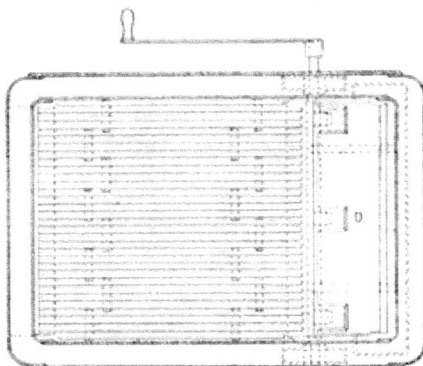

Fig. 65.   **Fig. 67.**

As much of the bituminous coal in this country, contains a great deal of material which causes it to clinker, or otherwise interferes with its free combustion, it has been found essential to provide locomotives with what are called shaking grates for " clearing the fire." A number of different grates of this kind which have been applied to locomotives at the Rogers Works are shown by the following engravings:

3

Figs. 68 to 71 represent the Allen & Hudson grate, which was patented by Mr. Albert J. Allen and William S. Hudson in 1858. The grate is composed of a series of cast-iron bars with lugs on their sides as shown in the plan. Underneath the bars are

Fig. 68.

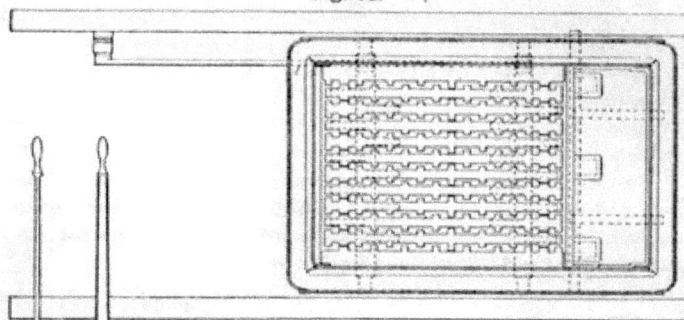

Fig. 69.

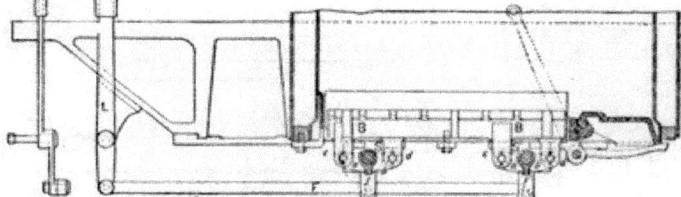

Fig. 70.

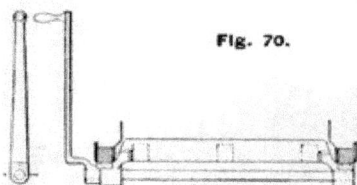

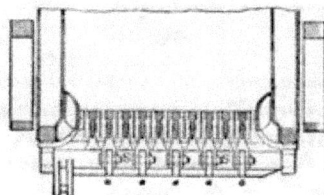

Fig. 71.

two cast iron rocking shafts, S, S', which have arms a, a' and b, b' on their opposite sides. Each grate bar has two projections c, c' and d, d', on its under side. To make it clear how the grate operates, it may be explained that the bar B, B, shown in Fig. 69, has the two projections c, c', attached to it, and that the projections d, d' are attached to the bar next to B, B. The projections c, c', are connected by pins to the arms a, a', and d, d', are attached to the arms b, b'. It is obvious then, that when the shafts S, S', are rocked, that the arms a, a' will rise, and b, b' will fall simultaneously, and vice versa, and that the grate bars connected to these arms will have a corresponding movement. As the alternate bars which compose the grate are connected to the arms on the right side of the shafts, S, S, and the bars between them are connected to the arms on the left side of the shafts, it is plain that the working of these shafts has the effect of giving a limited upward and downward movement to the bars in which each bar ascends as the next one on either side of it descends, and vice versa. This movement has the effect of breaking up the clinkers or other foreign or residuary matter that may collect upon the grate and which tends to choke the draft between the bars, and to cause such matter to work down between the bars into the ash pan, and also serves to evenly distribute the fuel over the grate.

The working of the shafts S, S', is effected by means of the lever L which is connected by a bar F, to vertical arms f, f', attached to the under side of the shafts. The grate is also provided with a drop door.

Fig. 72.

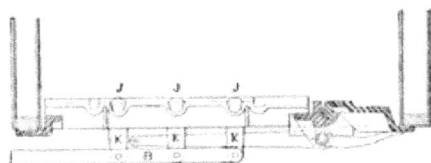

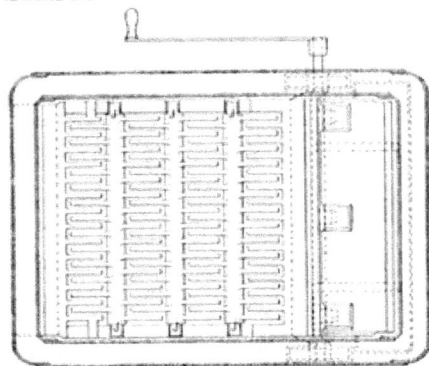

Fig. 73.

Figs. 72 and 73 represent what is called a "finger" grate, which consists of cast iron shafts, with projections or fingers on each side. These shafts rest in journals j, j, j', and are rocked by a lever (not shown in the engraving) and bar B, the latter connected to vertical arms K, K, K, attached to the shafts. It is obvious that as the shafts are rocked the fingers on one side rise, and those on the opposite side fall, and that the effect will be to

Fig. 74.

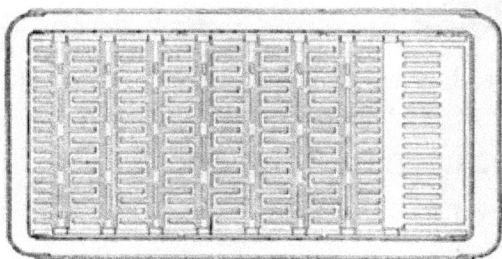

Fig. 75.

thoroughly shake up the fire.   Figs. 74 and 75 represent another form of finger grate.
Both the forms illustrated were first used in 1860.

Figs. 76 to 89 represent various forms of "rocking" grates as they are called.
These have transverse grate bars with journal bearings at each end, similar to those of

Fig. 76.

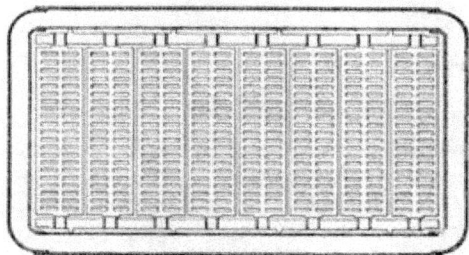

Fig. 77.

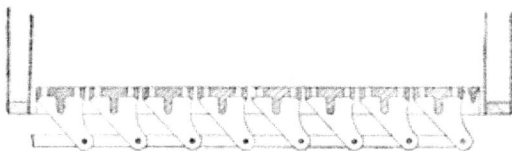

Fig. 78.

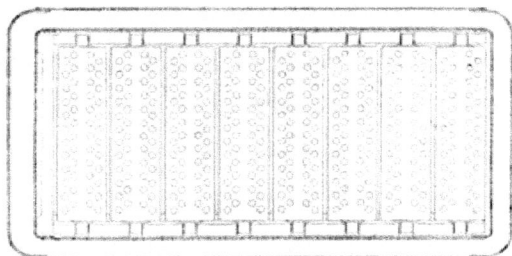

Fig. 79.

the finger grates. The bars are rocked on these journals, which has an effect similar to that of the finger grate in stirring up the fire. The construction and action of these grates will be obvious from the engravings.

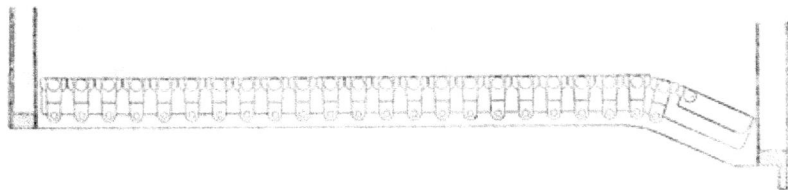

Fig. 80.

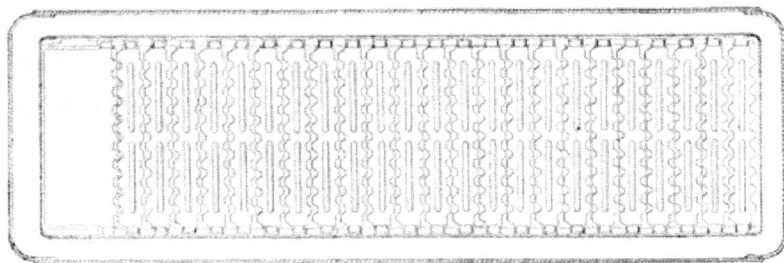

Fig. 81.

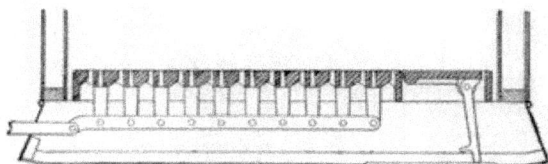

Fig. 82.

Fig. 83.

For burning anthracite coal the water tube grate is almost universally used. The form used on the Philadelphia & Reading railroad is shown in Figs. 30 and 31. The

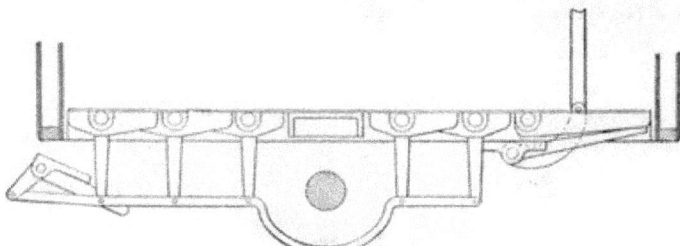

Fig. 84.

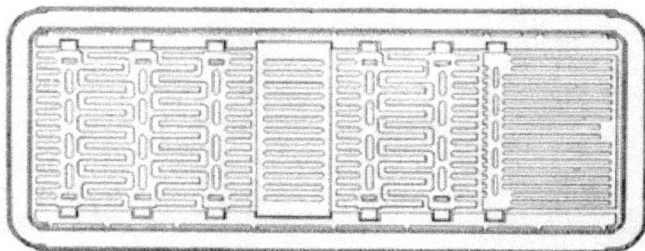

Fig. 85.

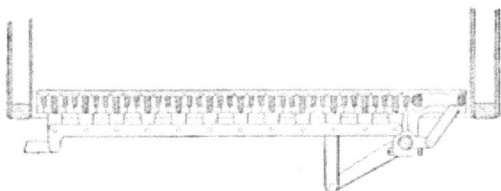

Fig. 86.

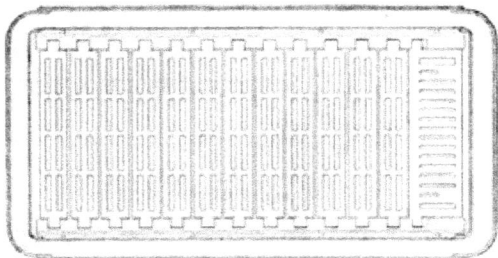

Fig. 87.

tubes are put in as shown in Fig. 31. Solid bars *B*, *B*, are substituted for every fourth tube. These bars pass through thimbles *T*, Fig. 30, in the back end of the fire-box, and can be drawn out through this thimble to clean or remove the fire.

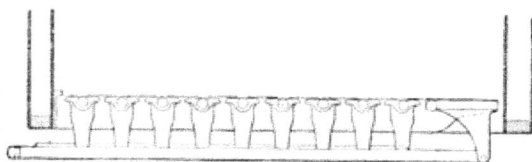

Fig. 88.

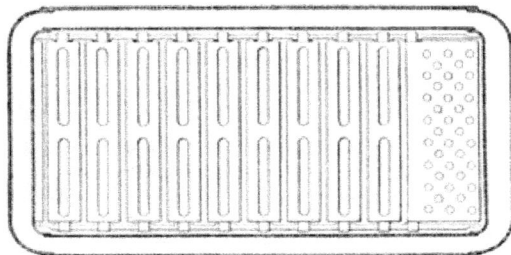

Fig. 89

Figs. 90, 91, and 92 represent a water grate recently introduced to burn bituminous coal.

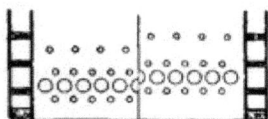

Fig. 91.

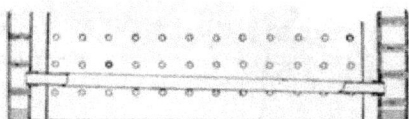

Fig. 90.

Fig. 92.

SMOKE BOXES.

As early as 1859 some engines were built at the Rogers Works for the New Jersey Railroad & Transportation Company with a form of extended smoke-box, shown in Figs. 93 and 94. A deflecting plate A was used in front of the top rows of tubes. In the same year the form of plate shown in Figs. 95 and 96, which had an adjustable piece B on its lower edge, was used on engines, both with and without the extended smoke-box. In 1862 the telescopic or adjustable petticoat pipe shown in Fig. 97 was applied to engines for the Nashville & Chattanooga Railroad. Figs. 98 and 99 show the extended smoke-box as recently applied to passenger engines. A, B, is a deflecting plate in front of the tubes, and C, C, C, is wire netting of number 13 wire, and $2\frac{1}{2}$ meshes to an inch. The exhaust nozzels F, F, it will be seen, are carried up above the horizontal centre line of the boiler. A receptacle D, for sparks, is attached to the under side of the smoke-box and has a sliding door E, for emptying the sparks and cinders which accumulate in the front end.

The extended smoke-box, when it was first introduced, met with little favor, but in recent years it has been extensively used.

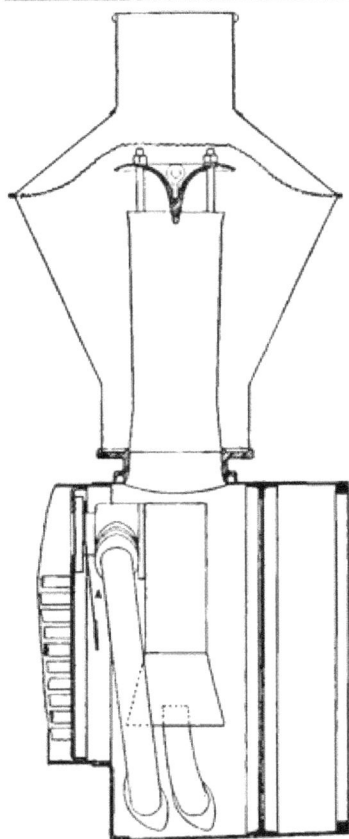

Fig. 93.

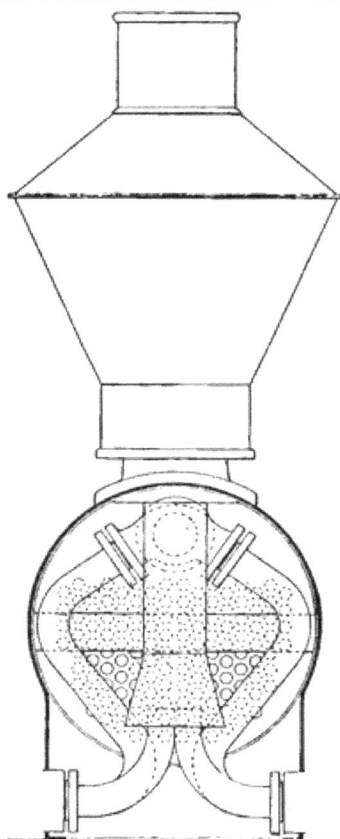

Fig. 94.

Fig. 95.

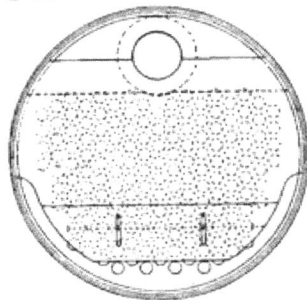

Fig. 96.

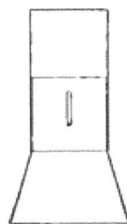

Fig. 97.

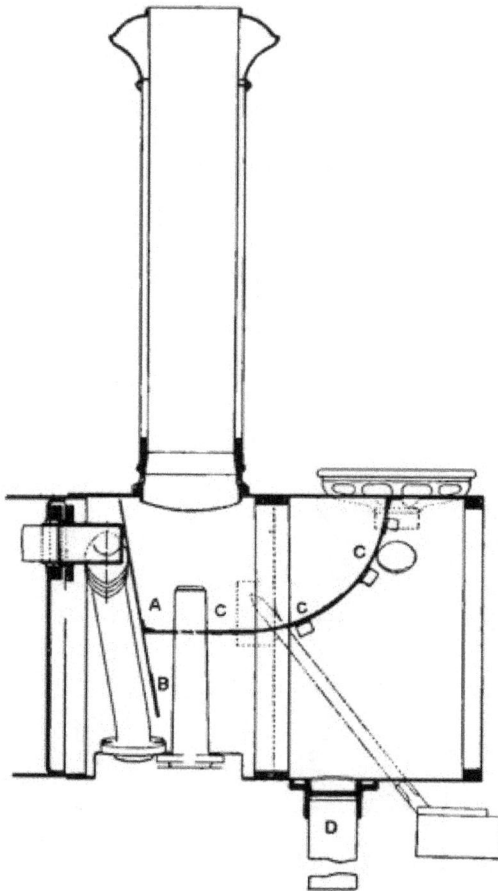
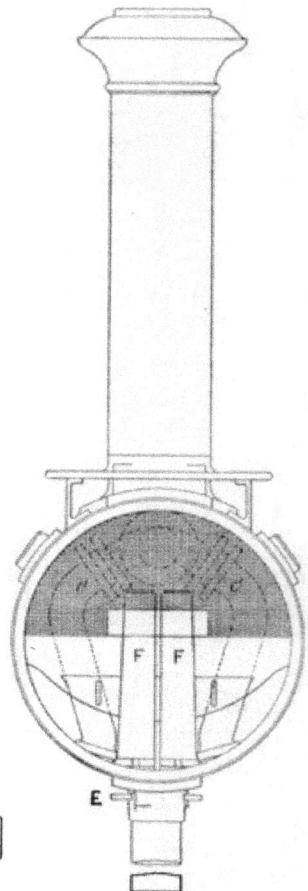

Fig. 98.   Fig. 99.

FEED WATER HEATER.

In 1859 Mr. Hudson designed a feed water heater, which is represented by Fig. 100, which he applied to a number of engines for the Southern Railroad of Chili, S. A. It consisted of a cylinder C, filled with small tubes F. At the end of the cylinder there was a chamber A and another B at the opposite end, which was connected together by

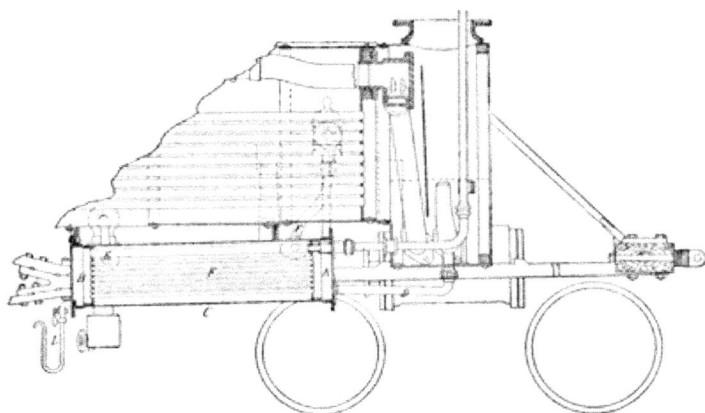

Fig. 100.

the small tubes. The exhaust steam was admitted to *A* from the exhaust pipes by a pipe *D*, and passed through the small tubes to *B*. The condensed water ran out through the pipe *L*, or it was conveyed to the ash pan. If not condensed, the steam passed through the pipe *G* to the chimney. The water from the pump entered the heater at *E*, and escaped by the pipe *F* to the check valve. This heater was used for some time, but as has occurred in numberless experiments with feed water heaters, it was finally abandoned under the impression that its cost was greater than the saving it effected.

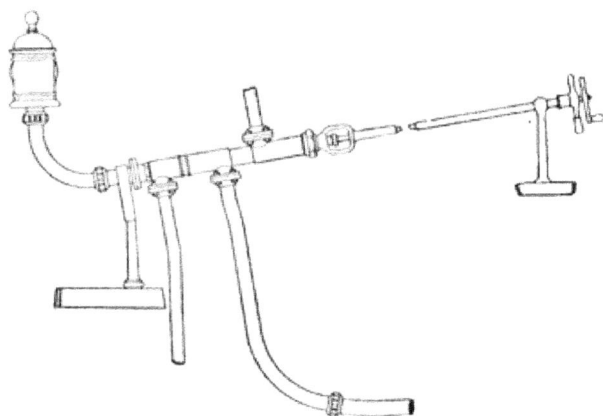

Fig. 101.

INJECTORS.

Injectors were first applied to locomotives at the Rogers Works in 1861. Fig. 101 shows the arrangement then used. Since that time they have been much improved and are almost universally used for feeding locomotive boilers.

SAFETY VALVES.

Figs. 102 to 108 represent different kinds of safety valves which have been used at various times, the construction of which is made sufficiently clear by the engravings, without other explanation. The dates when they were first used is given below each figure.

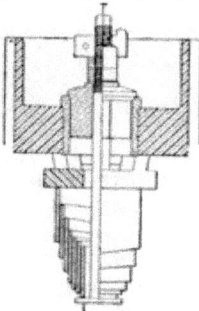
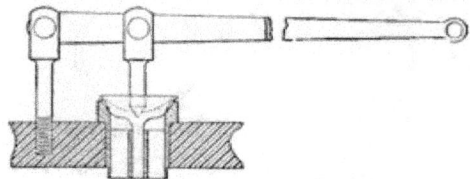

Fig. 103.
1872.

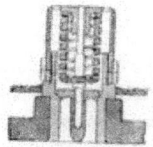

Fig. 102.
1861.

Fig. 107.
Steam Chest Safety
Valve, 1882.

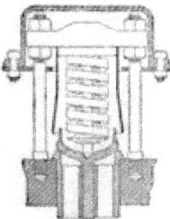
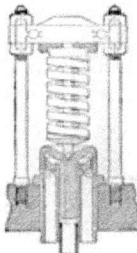
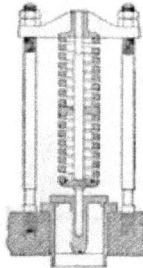
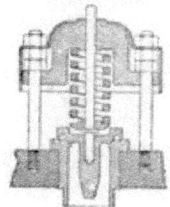

Fig. 104.
1872.

Fig. 105.
1875.

Fig. 106.
1882.

Fig. 108.
1883.

SMOKE STACKS AND SPARK ARRESTERS.

There is probably no part of a locomotive, unless it to be the valve gear, on which so much ingenuity has been exercised as on spark arresters. The very first engines built at the Rogers Works had some kind of bonnet or wire netting on the top of the chimney to "catch the sparks," and in the article on page 14 reprinted from the

American Railroad Journal, of December, 1839, it will be seen that at that time an inverted cone was placed on the "axis of the smoke-pipe to protect the wire gauze." Unfortunately there are no drawings extant of any of these early spark arresters. Figs. 109 to 137, however, give examples of later practice, and show different devices demanded by those who ordered locomotives of the Rogers Works. The date when they were first made and the fuel used is given under each of the figures.

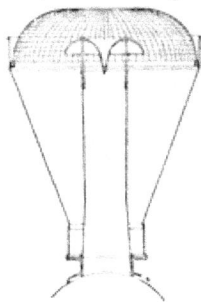

Fig. 109.
1854 —Wood.

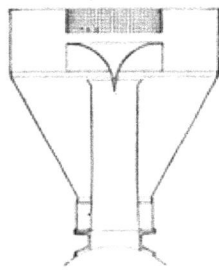

Fig. 110.
1854.—Wood.

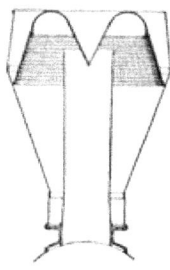

Fig. 111.
1854.—Wood.

Fig. 109 is what is called a bonnet stack, on account of the bonnet or hood of wire netting over the top. It was used for burning both wood and coal.

Fig. 110 had a deflecting cone and netting in the form of a cylinder over it.

Fig. 111 had a large deflecting cone with wire netting in conical form attached to the lower edge of the deflector.

Fig. 112 had a cone with flat horizontal netting of annular form around it.

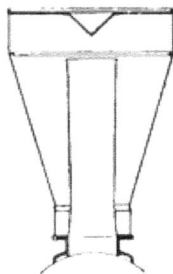

Fig. 112.
1856.—Wood.

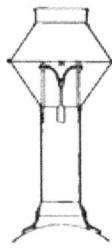

Fig. 113.
1858.—Wood and Coal.

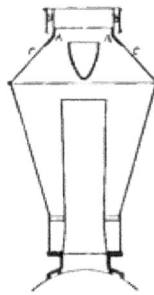

Fig. 114.
1860.—Bituminous Coal.

Fig. 113 is known as the diamond stack, from the form of the outline of its top. It had a deflecting cone, but no netting.

Fig. 114 had a curious shaped deflecting cone and a cast iron guard at *A*, *A*, to protect the sheet iron of the outside casing from the action of the cinders. It also had an annular opening *B B* around the top, the supposition being that the air coming in contact with the inclined surface C, C, would be deflected upwards through the opening B, B, and thus create an induced upward current out of the chimney.

Fig. 115 had a deflector with conical netting over it, which was open at the top.

Fig. 116 was the same as Fig. 115, but of different form.

Fig. 117 is a straight chimney with a cast iron grate at the top and a sliding damper at the base.

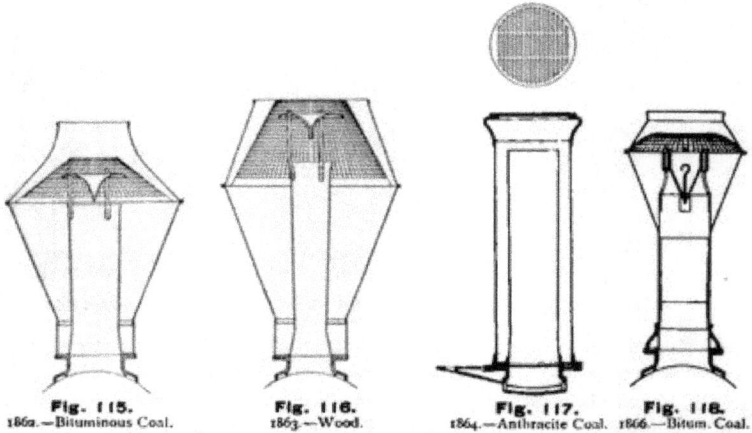

Fig. 115.
1862.—Bituminous Coal.

Fig. 116.
1863.—Wood.

Fig. 117.
1864.—Anthracite Coal.

Fig. 118.
1866.—Bitum. Coal.

Fig. 118 had a deflector with netting over it, which was open in the middle. The opening was surrounded by a cylindrical shaped netting as shown.

Fig. 119 was the same as Fig. 110, but of different shape and proportions.

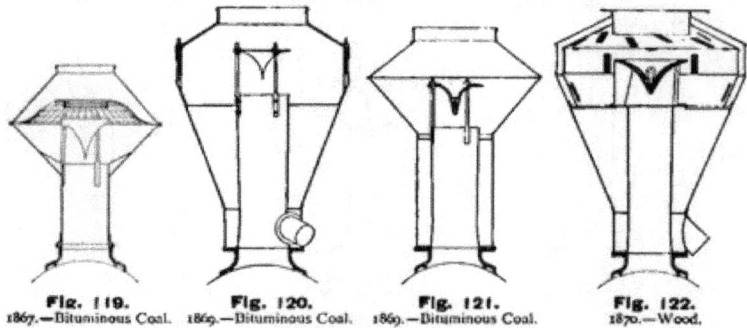

Fig. 119.
1867.—Bituminous Coal.

Fig. 120.
1869.—Bituminous Coal.

Fig. 121.
1869.—Bituminous Coal.

Fig. 122.
1870.—Wood.

Fig. 120 had a deflector with a very large casing or receptacle for sparks.

In Fig. 121 the netting was placed horizontally over the deflector.

Fig. 122 represents the celebrated Radley & Hunter stack, which was at one time very generally used for wood burning locomotives.

Fig. 123 has a conical shaped netting over the deflector, with an opening in the centre surrounded by another netting of cylindrical shape.

Fig. 124 has a deflector with a wire netting bonnet over it.

Fig. 125 is similar to Fig. 124.

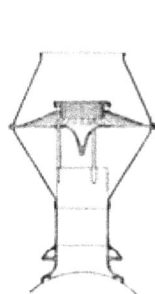

Fig. 123.
1872.—Wood and Coal.

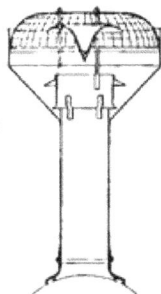

Fig. 124.
1872.—Wood.

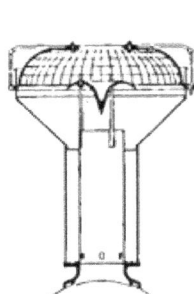

Fig. 125.
1872.—Coal.

Fig. 126.
1873.—Coal.

Fig. 126 has a deflector with a circular opening above it, and cylindrical guard around the edge made of perforated sheet iron or copper.

Fig. 127 shows what is called a "straight" stack, and has no spark arresting attachments.

Fig. 127.
1879.—Bitum. Coal.

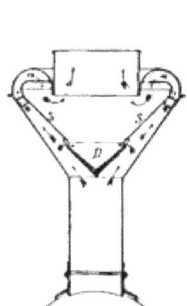

Fig. 128.
1879.—Bitum. Coal.

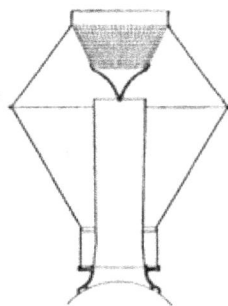

Fig. 129.
1879.—Bituminous Coal.

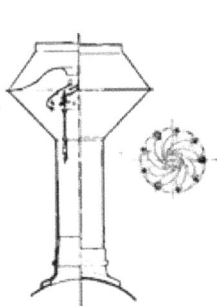

Fig. 130.
1881.—Bituminous Coal.

Fig. 128 represents the Fontaine stack. This has a deflector *D*, to which a shield *S*, *S*, is attached. Between the shield and the outer casing there is space for the passage of the products of combustion, which escape in the direction indicated by the darts.

Fig. 129 has an outside case or receptacle for sparks which was unusually large. It had a deflector surmounted with an inverted cone of wire netting. This forms a guard for the opening at the top so that all the smoke must pass through the netting to escape into the open air.

Fig. 130 shows a stack with a spark arrester patented by Wm. S. Hudson in 1877.

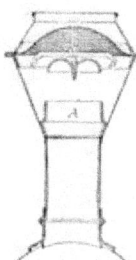

Fig. 131.
1881.—Bituminous
Coal.

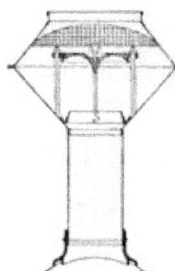

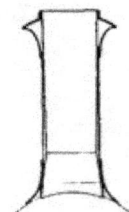

The reflector is formed of what Mr. Hudson described as "peculiarly curved screw blades," which are shown on plan in the engraving. "The gaseous products of combustion," the inventor says in his specification, "mingled with more or less small masses of coal in various conditions, are thrown violently upward through the cylindrical chimney, and, striking in the hollow interior of the dome-like set of wings, are thrown into a spiral motion without completely interrupting their upward motion. The solid matter is projected against the wire netting. A portion of the gaseous matter follows the same course, and another portion moves inward, and, passing freely upwards through the open space in the centre."

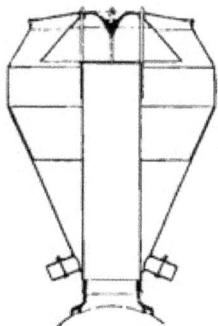

Fig. 132.   Fig. 133.   Fig. 134.   Fig. 135.
1881.—Bituminous Coal.   1882.—Bituminous Coal.   1882.—Bituminous Coal.   1882.—Bitumin's Coal.

Fig. 131 is provided with a casting A, which forms what was called a stricture for some purpose not clearly understood. The usual deflector was suspended from a casting B, B, with radial arms meeting in the centre.

Fig. 136.
1882.—Bitumin-
uos Coal.

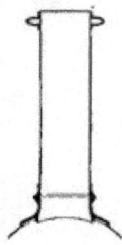

Fig. 132. This stack had a large receptacle for sparks, with a deflector placed at the top. The latter had a sheet iron guard around the edge, as shown in the engraving. The top of the stack was open; no netting was used.

Fig. 133 had a deflector with wire netting over it as shown.

Fig. 134 was similar to Fig. 133, but of somewhat different proportions. It also had what was called a "stricture" or contraction of the opening at S. The effect of this was to concentrate the escaping current and cause the sparks to impinge directly against the deflector.

Fig. 135 represents what is called a "straight" stack without spark arrester of any kind.

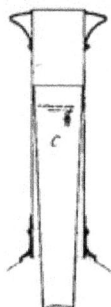

Fig. 137.
1882.—Anthracite
Coal.

Fig. 136 illustrates a straight stack with a long inverted cone inside of it. This was made of perforated sheet iron, and was connected at the bottom to the exhaust pipe, so that they discharged inside of the cone and the smoke had to pass through the perforations in the inverted cone. The perforations were $1 \times \frac{3}{8}$ in.

Fig. 137 shows a straight stack for anthracite coal.

Fig. 138.

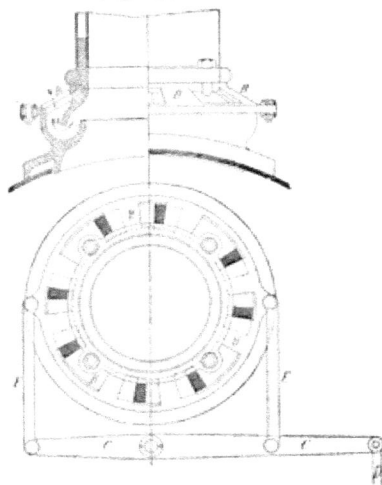

Fig. 139.

CHIMNEY DAMPERS.

Figs. 138 and 139 represent a form of damper recently devised and patented in 1885 by Mr. H. A. Luttgens, who has been the chief draftsman in the Rogers Works for 28 years past. It is intended for the chimneys of coal burning engines. Its object is to diminish the effect of the exhaust by admitting air at the base of the chimney, and thus obviating the necessity for opening the fire door and admitting cold air into the fire-box.

In constructing the damper the base of the chimney is made of the form shown in half section on the left side of Fig. 138, from which it will be seen that there are cavities A, through which air is admitted, as indicated by the darts. The outer openings of these cavities are shown by the dark shading and dotted lines in the plan, Fig. 139. On top of these openings is a circular valve or cover with openings corresponding to those in the base of the chimney. This valve by being turned a part of a revolution by means of the links E, E, and lever C, C, which is connected with the cab by a rod D, will cover or uncover the openings leading to the cavities in the base of the chimneys, and thus air may be admitted to or shut off from the chimney at pleasure.

4

THE ENGINES.

CYLINDERS.

The first method of fastening outside cylinders was to bolt them to the smoke-box, which was made of sheet or plate iron. When the cylinders were steeply inclined, as shown in Fig. 17, page 17. This could be done without difficulty, but when they were placed lower down it was necessary to extend the smoke-box downward. The lower part was usually made rectangular in shape, as shown in Fig. 140, with a heavy wrought iron bar *B, B, B,* riveted around the inside at the front end. The cylinders were

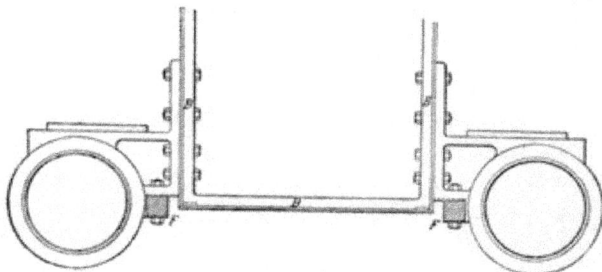

Fig. 140.

then bolted to the outside of the smoke-box and to the frames *F, F,* as shown in the engraving. This method of fastening was first used in 1844.

Inside cylinders were attached to the smoke-box and frames as shown in Fig. 141.

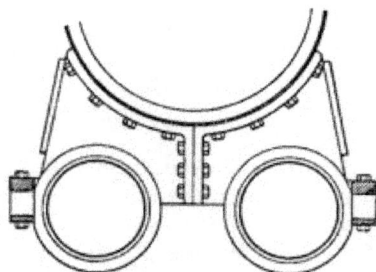

Fig. 141.

The next step, which was taken in 1853, was to make the bottom *B, B,* Fig. 142, of the smoke-box of a heavy wrought-iron plate. This extended outward so as to rest on top of the frames *F, F.* The cylinders were then placed on top of the plate and bolted to it, and to the smoke-box and frames, as shown. A bar *C, C,* with T ends was also placed crosswise between the bar *B, B,* to keep it apart and stiffen the whole attachment.

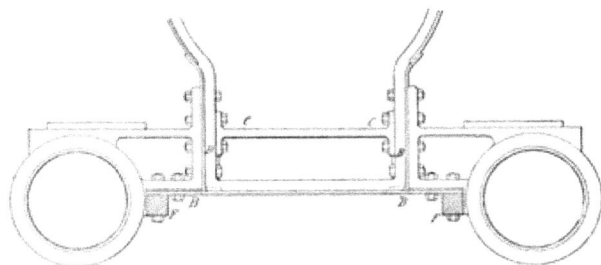

Fig. 142.

In 1865 the arrangement shown in Fig. 143 was adopted. The smoke-box in this case was substantially like that shown in Fig. 142, but a cast-iron bed E, E, was

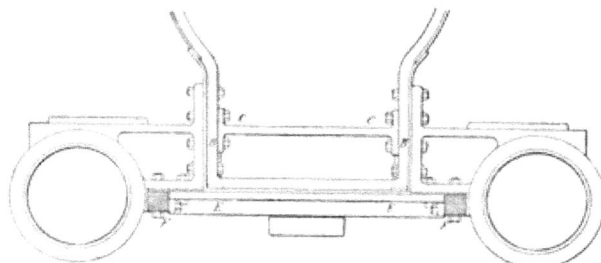

Fig. 143.

placed between the two frames F, F, and bolted to them by flanges. The smoke-box

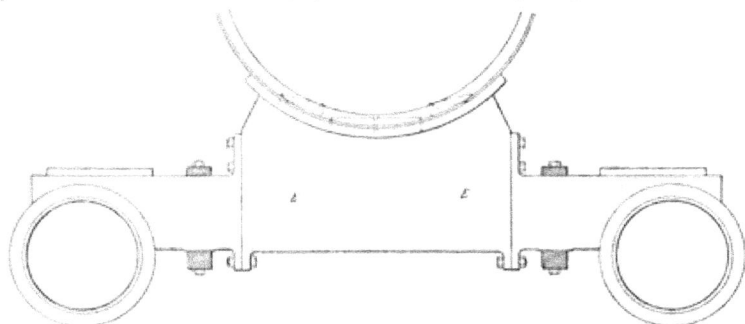

Fig. 144.

was then placed on top of the bed plate and bolted to it. The cylinders were bolted to the bed plate frame and smoke box as shown.

About the same time the plan represented in Fig. 144 was put in use. In this the smoke-box was made cylindrical and a heavy bed casting E, E, with steam and exhaust pipes cast in it, was bolted to it by suitable flanges. The cylinders were then attached to the frames and to this casting as shown.

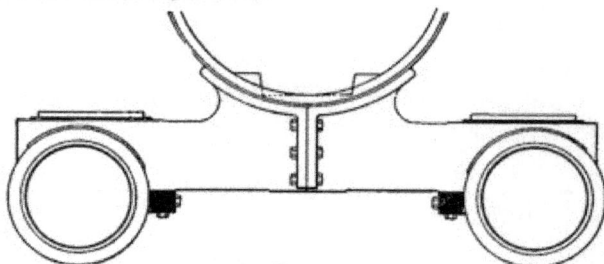

Fig. 145.

In 1871, the plan shown in Fig. 145 was adopted. The smoke-box was cylindrical, and one-half the bed casting was cast with each cylinder. They are bolted together in the centre as shown. This plan is now almost universally used in this country and makes a very neat, strong, and satisfactory job.

VALVES AND VALVE GEARING.

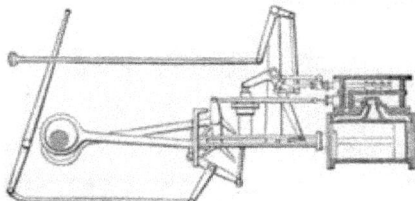

Fig. 146.

The main valves which were first built by Mr. Rogers were of the ordinary D pattern and the valve-gearing was a form of hook motion. In some cases as shown in Fig. 14, the eccentrics were outside of the journals and wheels. Unfortunately, there are no authentic drawings in existence of the various forms of valve gearing which were at first used. At an early date Mr. Rogers was impressed with the importance of using steam expansively, and in 1843 and 1846 he

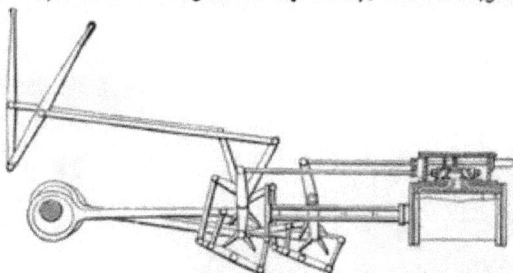

Fig. 147.

designed and used the valve gearing shown in Fig. 146. It serves to show the thought he was giving at that date to the subject of working steam expansively.

Fig. 147 shows another plan which he introduced in 1847.

When the link-motion was introduced into this country its use was violently opposed by many locomotive builders and master mechanics. Mr. Rogers was one of

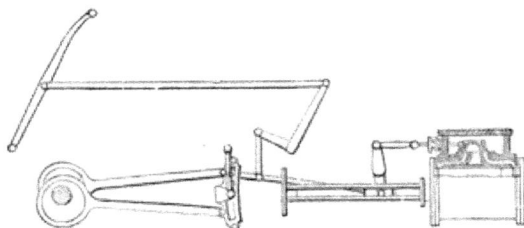

Fig. 148

the first American engineers to recognize its merits. In 1849 he used the suspended link-motion, shown in Fig. 148, for some engines for the Hudson River Railroad, and in

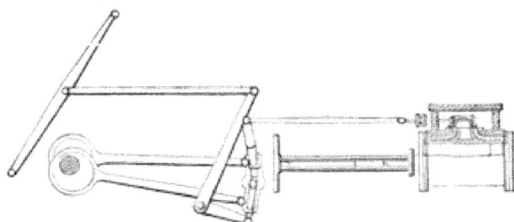

Fig. 149.

1850 he applied the shifting link motion, shown in Fig. 149, to some engines which he built. It will be noticed that in this case the lifting-shaft was below the link. In the

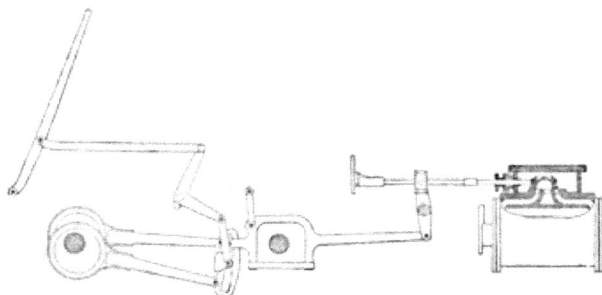

Fig. 150.

same year he designed the form of link-motion shown by Fig. 150 for some ten-wheel engines, the front wheels and axles of which came in the way of the rocking shaft. In this case the lifting shaft was above the link.

Fig. 151 represents a combination of link motion with an independent graduated cut-off valve. It was used on several locomotives built at the Rogers Works in 1854, and it is said was found to be beneficial in economizing fuel.

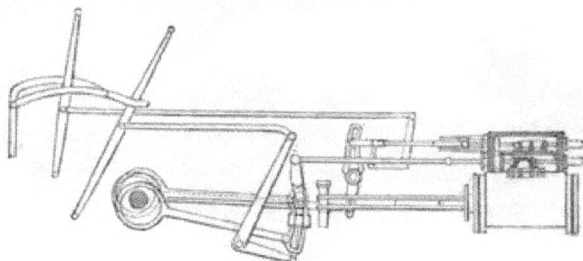

Fig. 151.

For many years the form of valve-gear, shown in Fig. 149, was used by Mr. Rogers, and after his death it was applied to many engines; but in 1862 Mr. Hudson

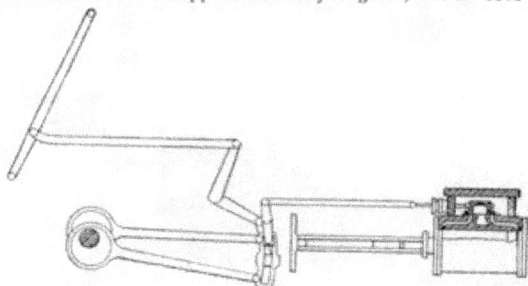

Fig. 152.

designed the form of link-motion shown by Fig. 152, in which the lifting shaft was placed above the link. This is the form which is now most commonly used. The link motion

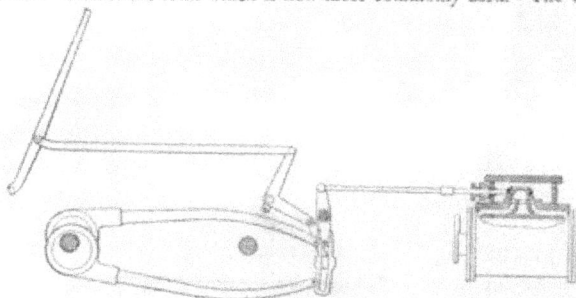

Fig. 153.

shown by Fig. 153 was also designed the same year by Mr. Hudson and applied to some ten-wheel engines, in which the front wheels and axle came in the way of the rocking shaft.

In 1866 the valve gearing shown in Fig. 154, which was designed and patented by Messrs. Uhry & Luttgens, was applied to an engine for the Central Railroad of New Jersey. In this there is an ordinary shifting link worked by two eccentrics and connected with a pin attached the lower arm of a rocking shaft in the usual way. What may be called a supplementary rocking-shaft R, R', was pivoted to the top pin of the main rocking-shaft. The lower arm R' of the supplementary rocking-shaft is bent into a half circle, as shown, in order to clear the main rocking-shaft M. The supplementary rocker is worked by a cam, O', which was connected to a pin L. The effect of the action of the cam is to accelerate

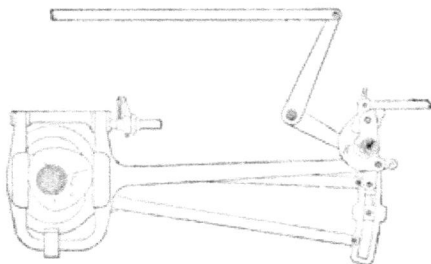

Fig. 154.

the movement of the valve at the time that it opens the ports for admission and exhaust. Its adjustment is the same as that of the link-motion, and at the higher grades of expansion it gives about 50 per cent. greater opening of steam port. The point of exhaust is retarded from 5 to 6 inches beyond the link-motion, while the point of compression

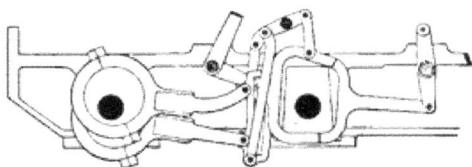

Fig. 155.

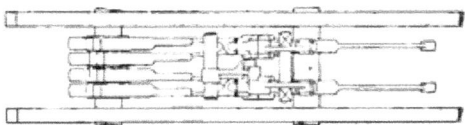

Fig. 156.

remains the same. The size of opening of the exhaust port is somewhat larger than with the link-motion, and it is opened in less time, thereby producing a strong and clear exhaust.

Its objectionable feature is the cam as a mechanical device for locomotives. Whether this objection would be as great if used with a balanced valve as it is with an ordinary slide-valve remains yet to be proved.

Figs. 155 and 156 shows the methods which was adopted in 1873, in applying the Allen link-motion to some narrow gauge engines for the Patillas Railway, S. A., in which the front axle was in the way. Ordinarily the Allen link is made straight, but in this case Mr Hudson found that it would not give a satisfactory movement to the valve without curving the link slightly.

Fig. 157 shows another method of applying a link-motion to engines in which the front axle was in the way. This was used in 1881.

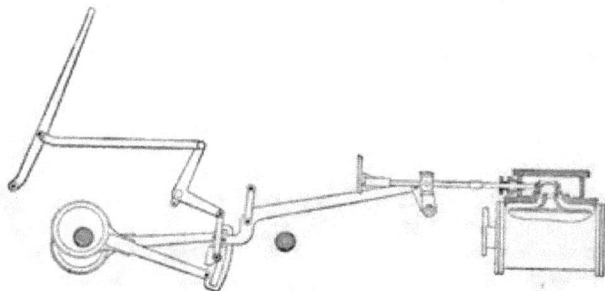

Fig. 157.

COUNTERWEIGHTS FOR LINKS.

When shifting links were introduced it became important to counterbalance their weight so as to lessen the effort required to move them. The arrangement shown in Fig. 158 was adopted in 1858. In this the counterweight W was attached to an arm or bell crank forged on the reversing lever.

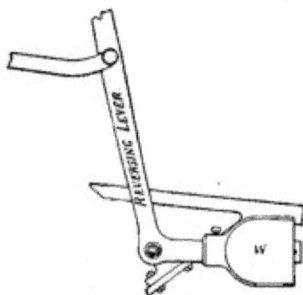

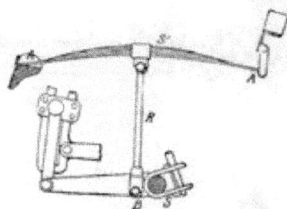

Fig. 158.   Fig. 159.

The unwieldy character of a counterweight led to the substitution of springs of various forms. The plan shown in Fig. 159 was adopted in 1859. In this a half elliptic spring S' which was attached by its ends, A, A, to fixed parts of the engine, was connected by a rod R to a short arm B which was keyed on the lifting-shaft by a strap S, as shown.

Another plan of applying a semi-elliptic spring is shown in plan in Fig. 160. In this case the spring S was connected to a short arm B forged on the middle of the lifting shaft.

In 1860 a spiral spring, Fig. 161 and 162, was used. The inner end of this spring was attached to the lifting-shaft S' and the other end was fastened to a case in which it was enclosed. The case was prevented from turning by a bolt B. The required amount of tension was brought on the spring by turning the case, and the bolt was adjusted in any one of the holes, which were arranged in a circle as shown in the engraving.

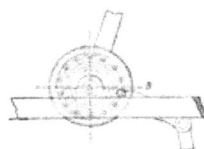

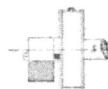

Fig. 161.   **Fig. 162.**

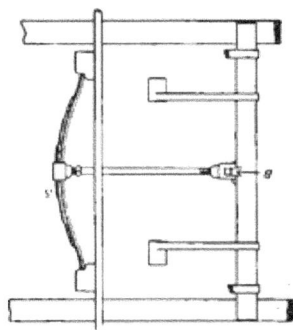

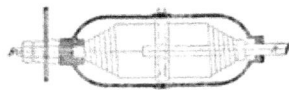

Fig. 160.   **Fig. 163.**

In 1873 a pair of volute springs was substituted for the semi-elliptic spring. These volute springs are shown in Fig. 163. They were inclosed in a case and fastened by a bolt B to one of the cross beams, and were connected by a rod R to a short arm on the lifting-shaft, like that shown in Fig. 160. In this instance the rod R was subjected to a compressive strain by the tension of the two volute springs.

Fig. 164 shows a helical spring which was applied in 1875 for the same purpose. This was also enclosed in a cylindrical case, which was fastened to a fixed part of the engine. A chain C, C, was fastened at one end to the shaft, and wound around it as shown. The other end was attached to a rod R which was screwed into a collar K. When the shaft was turned the spring was com-

Fig. 164.

pressed. Its tension could be adjusted by means of the screw end on the rod so as to balance the weight of the link.

SLIDE VALVES.

The first slide valves used at the Rogers Works were the ordinary D pattern. In 1853 Mr. Rogers adopted the Hackworth valve, Fig. 165, with double exhaust ports.

This valve had about $\frac{3}{16}$ in. lap at a, a', and only $\frac{1}{32}$ at b. Consequently the steam was not released at a, a', as shown in Fig. 166, until the steam port B was opened nearly $\frac{3}{16}$ in. wide at b. Then the two ports a and a' each commence to open. The exhaust was thus delayed, but when it did begin the steam escaped through both of the openings at a, a'. The area of the exhaust opening was therefore doubled when the release occurred. This

Fig. 165.

Fig. 166.

form of valve was used up to 1872 and applied to more than 250 engines, but its advantages did not seem to compensate for the increase in its area, which was due to the double ports.

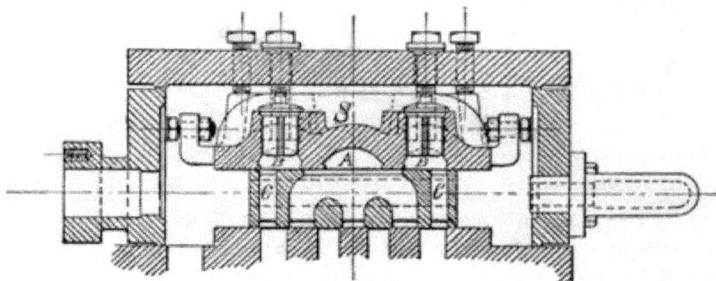

Fig. 167.

In 1864 Mr. John Gleason patented a valve which Mr. Hudson afterwards modified and introduced in the form shown by Figs. 167 and 168. This had a saddle S

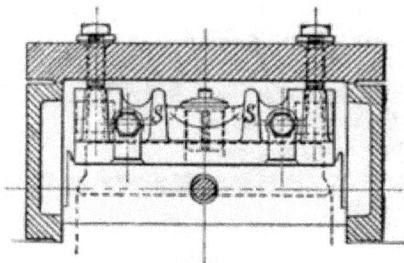

Fig. 168.

on top, the position of which was regulated by set screws, as shown. The saddle had steam openings B, B, and an exhaust opening A on its under side. The valve had double exhaust ports the same as are shown in Figs. 165 and 166. In addition it had two supplementary steam passages C, C. In the position shown in Fig. 169, not only was the steam port B open at b, but there was another opening at a through which steam passed to the supplementary port c, as shown by the dart, and thence to the cylinder. The opening of the steam ports was thus doubled during the early portion of the period of admission. A similar action occurred on the exhaust side. This valve was tried, but with rather doubtful resulting advantages.

Fig. 169.

In 1868 the Bristol roller slide valve, shown by Figs. 170 and 171, was applied to a number of engines. This valve rested on a series of rollers R, R, placed in each side of the valve. They were connected to a frame F, F, their axles or spindles having a little play in their journals. Steel plates were attached to the valve on each side, and others to the valve-seat, so that the rollers rested on the latter below, and the valve was

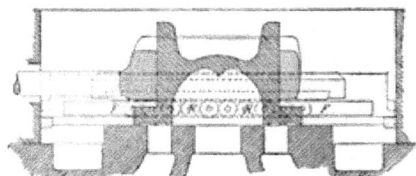 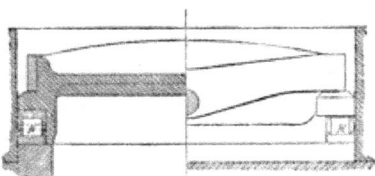

Fig. 170.   Fig. 171.

carried by the upper plates, which in turn rested on the rollers. With careful workmanship, the pressure of the valve could be carried on the rollers, and as it wore, of course, there was little or no contact between its face and seat. These valves were quite extensively introduced, but their use has been gradually abandoned.

In 1882 two forms of the Allen valve were introduced. Figs. 172 and 173 shows an Allen valve with Richardson's "balanced" or equilibrium device applied to it, and Fig. 172 shows an Allen valve with extensions to increase its length, and with steam-ports to admit live steam from below into the supplementary port S, S. The Allen valve although an American invention, was not used on locomotives in this country to any extent until after the expiration of the patent on it. It is now extensively used and its advantages are generally recognized.

Fig. 172.

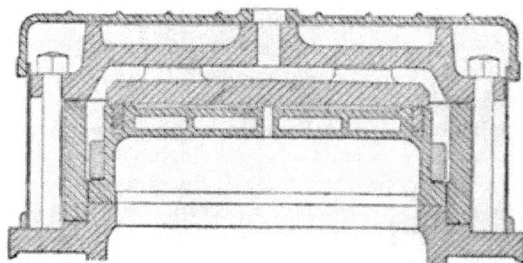

Fig. 173.

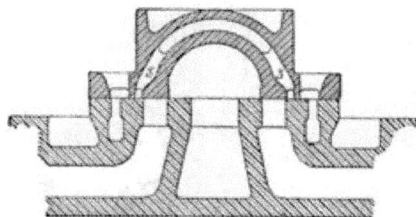

Fig. 174.

THE RUNNING GEAR.

FRAMES.

The frames used on the first locomotives built by Mr. Rogers (see Figs. 12 and 14), were made of two plates, with wood filling between them. The journal bearings were outside the wheels, as shown in the Figs. referred to.

Bury, who first introduced the hemispherical topped furnace in England also used bar frames on some of his engines. It seems probable that his form of fire-box and method of constructing frames were simultaneously introduced here. There are no drawings extant of the early frames made at the Rogers Works, but in 1844 the form of frame shown in Fig. 175 was used. It consisted, as will be seen, of a straight bar on

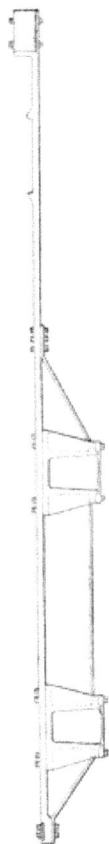

Fig. 175.

Fig. 176.

Fig. 177.

Fig. 178.

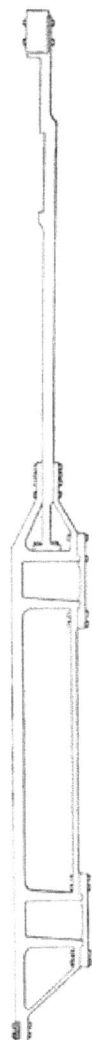

Fig. 179.

top, with cast-iron pedestals bolted to it and braced at the bottom very much after the manner in use at present.

In 1850, wrought-iron pedestals were substituted for those of cast-iron, as shown in Fig. 176. In 1854, the whole frame was forged in one piece, as shown in Fig. 177. With this form of construction some difficulty was encountered in cases of collision, and other accidents to locomotives, in which either the front or the back ends of the frames were injured. It then became necessary to take down the whole frame to repair one end. This led to making the front and back ends in separate pieces and bolting them together, as shown in Fig. 178. With this plan, if either end was taken down it was necessary to remove one pair of driving wheels. As the front part of the frame is usually injured in accidents, it was desirable to be able to take it down without removing any of the driving wheels. The plan shown in Fig. 179 was therefore adopted in 1868. In this the front end is bolted to the back end, ahead of the front pedestals, so that the front part can be removed without disturbing the driving wheels, if it is desirable to do so. This form of construction is the one which is still used and has been very generally adopted on American locomotives.

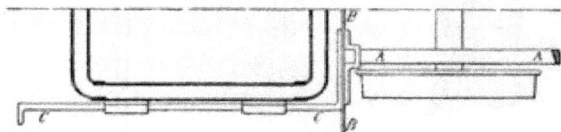

Fig. 180.

One of the difficulties in the construction of narrow guage engines is that there is not room enough between the frames for the fire-box, and it must therefore be made very narrow. To obviate this Mr. Hudson in 1873 designed the frames shown in Figs. 180 and 181. In this plan the main frames A, A, are placed in the usual position inside of

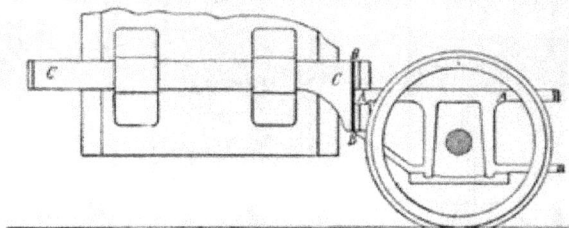

Fig. 181.

the wheels. A cross plate B, B, which projected outside of the wheels, was bolted to the back ends of the frames. Two flat bars C, C, were then bolted to the cross-plate, and placed far enough apart so as to give sufficient room between them, for a fire-box of the width required. The tank locomotive represented by plate XXIII has a frame of a similar plan.

SPRINGS AND EQUALIZING LEVERS.

Ordinary equalizing levers were used between the driving-wheels on the engine

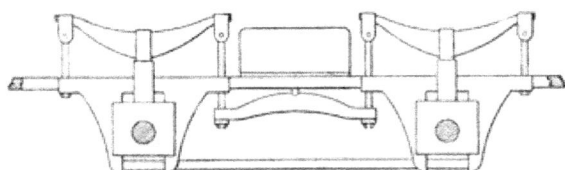

Fig. 182.
1837.

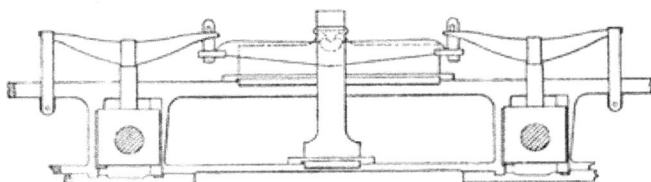

Fig. 183.
1850.

represented by Fig. 18, which was built in 1845. Mr. Rogers appreciated their value, and very few if any engines were afterwards built without using them in some form.

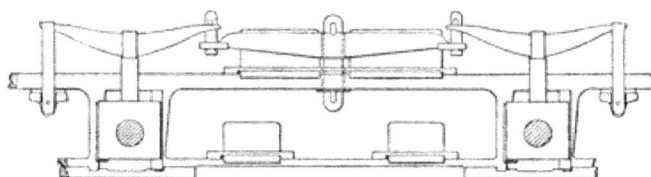

Fig. 184.
1860.

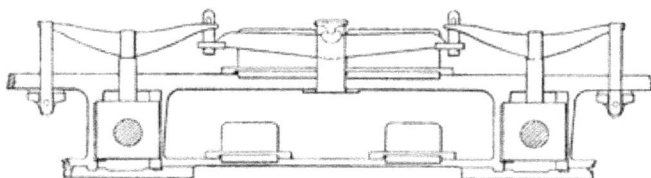

Fig 185.
1867.

Figs. 182 to 186 show the forms of spring and equalizing lever arrangement that were successively used for eight-wheeled American engines.

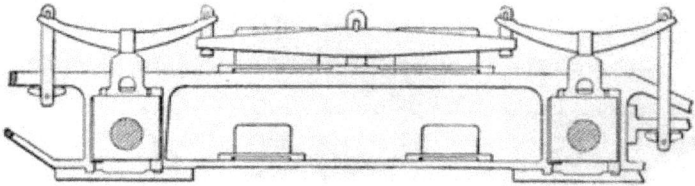

Fig. 186.
1880.

Figs. 187 and 188 represent a plan adopted for narrow guage engines in 1878. The purpose was to allow a wider fire-box to be used than is possible when the springs are placed alongside of it.

Fig. 187.

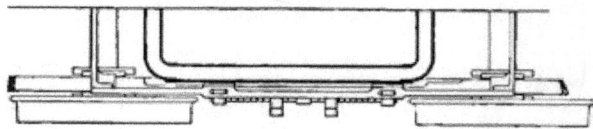

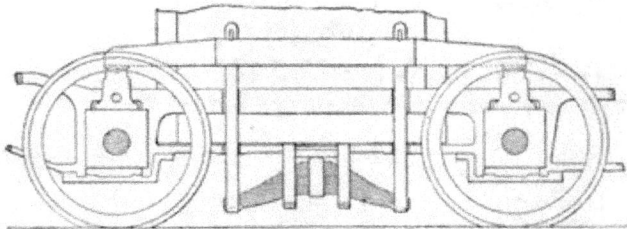

Fig. 188.

Fig. 189 shows the arrangement of springs used in 1880 for consolidation engines. The springs for the front axle are not shown in the engraving. Their connection with

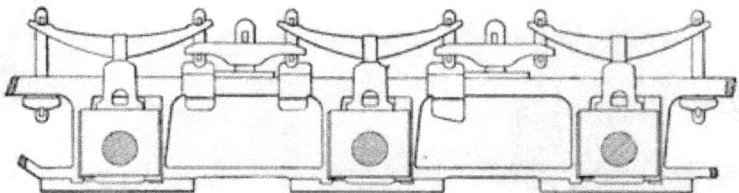

Fig. 189.

the leading truck and other applications of equalizing levers will be described farther on under the head of trucks.

Robt. S. Hughes

DRIVING WHEELS.

A method of constructing driving-wheels for 5 ft. gauge roads, which it is expected will have their guage changed and which will therefore require to have their tires brought nearer together to conform to the altered guage, is shown in Figs. 190 and 191. A projection P, P, is cast on the inside of the wheel centre. The tires are then set to conform to the wide guage. When the time comes to narrow it the tires are simply moved farther in. The projection of the wheel centre which is left on the outside is then turned off, which leaves the wheel in proper condition for the narrow guage. The first engine with wheels constructed in this way was for the Alabama & Great Southern Railroad in 1881. Since then all engines built for 5 ft. guage roads, which it is expected will be changed to the standard guage are made in this way.

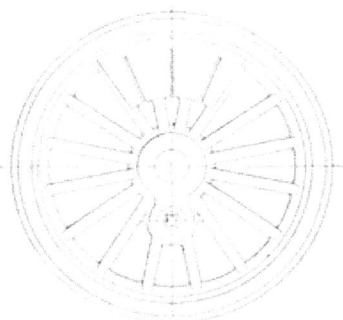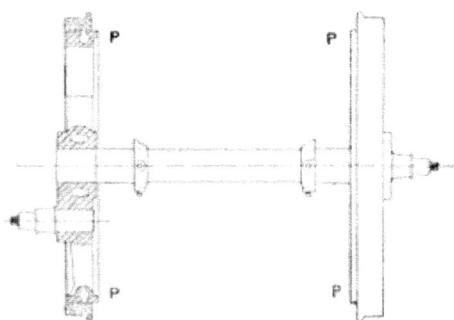

Fig. 190.　　　　　　　　　　　　Fig. 191.

In addition to the extra depth of the rim of the wheel centre the spokes are extended on the outside so as to form a brace or support to the projecting rim. These braces as well as the projection are turned off when the guage of the wheels is narrowed.

This expedient for changing the guage was suggested by Mr. James Cullen, Master Mechanic of the Nashville, Chattanooga & St. Louis Railroad, to Mr. R. S. Hughes, Secretary and Treasurer of the Rogers Locomotive and Machine Works. The plan was at once adopted for engines for the 5 ft. gauge.

CONNECTING RODS.

Figs. 192 to 203 represent various forms of connecting rods which have been made at the Rogers Works at various times. The dates when they were first used are appended to the engravings, which show the construction so clearly that further description is not needed.

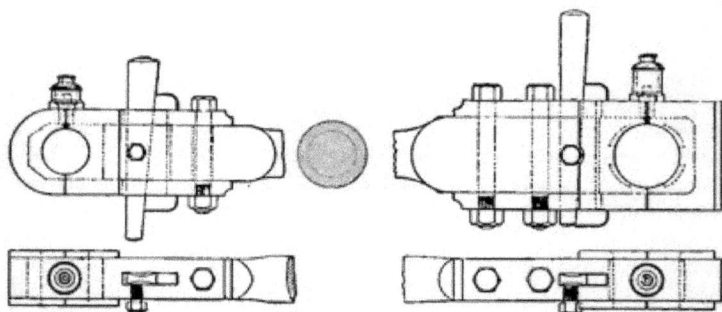

Fig. 192.
1837.

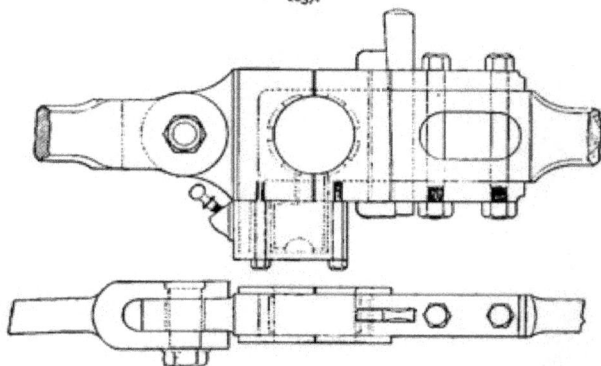

Fig. 193.
1845.

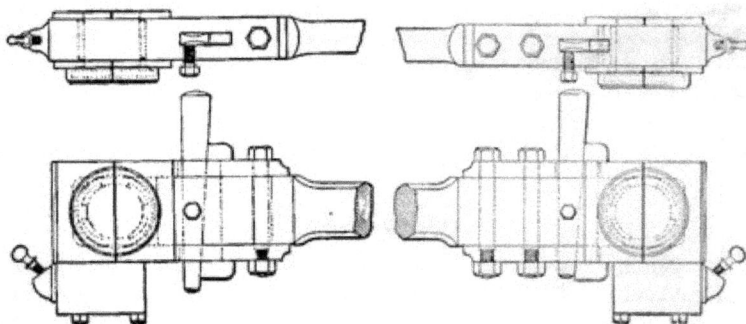

Fig. 194.
1854.

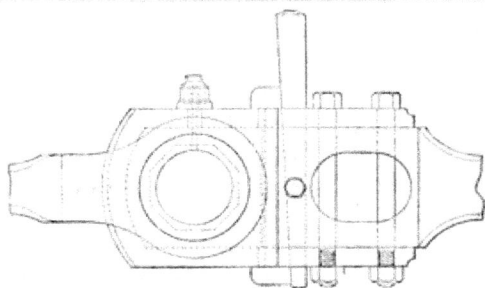

Fig. 195.
1861.

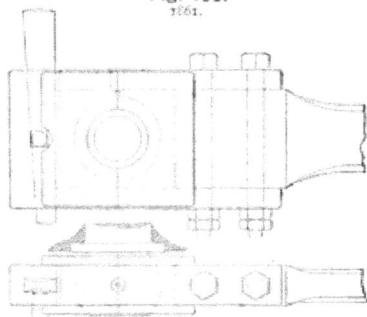

Fig. 196.
1870.

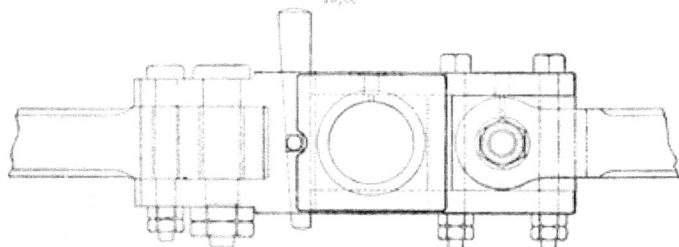
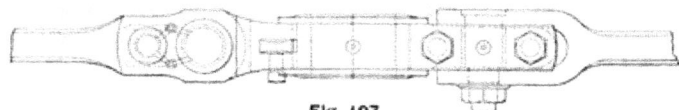

Fig. 197.
1870.

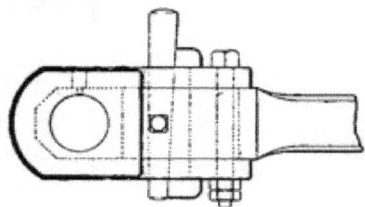

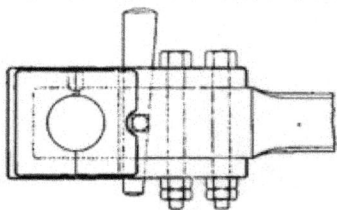

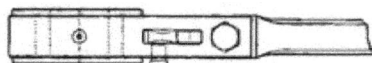

Fig. 198.
1880.

Fig. 199.
1880.

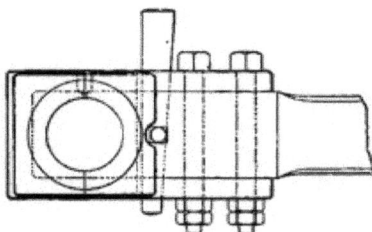

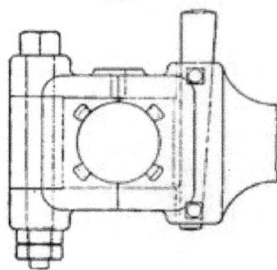

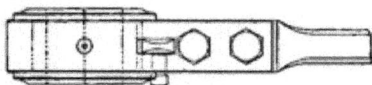

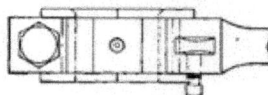

Fig. 200.
1880.

Fig. 202.
1882.

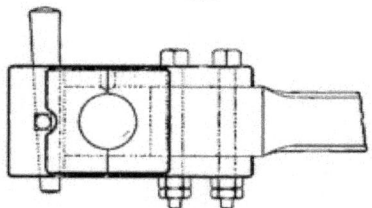

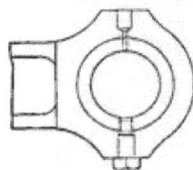

Fig. 201.
1880.

Fig. 203.
1885.

TRUCKS.

When trucks were first used in this country, it was considered very essential that their axles should be as near together as possible, and from Figs. 12 to 22 it will be seen that the trucks of all the early engines built at the Rogers Works had their wheels as close to each other as they could be placed. With outside cylinders this could be done without difficulty so long as the cylinders were inclined, but owing to the rolling motion which was produced by cylinders with a steep inclination, and also other inconveniences, the tendency was to lower the cylinders, and, excepting with large driving wheels, this made it necessary to spread the truck wheels farther apart. Finally the cylinders were brought down horizontal, and it was then found that there was really no disadvantage in placing the wheels the required distance apart, but rather the reverse.

Excepting as they are shown in the small engravings of the engines, no drawings of the early trucks which were made at the Rogers Works have survived to the present time.

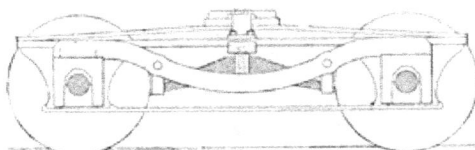

Fig. 204.
1850.

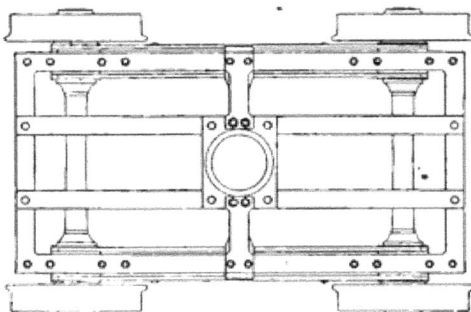

Fig. 205.

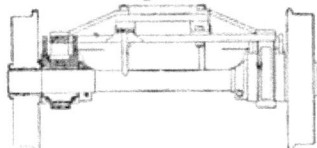

Fig. 206.

In 1850 Mr. Rogers designed the truck shown by Figs. 204, 205, and 206. This had a rectangular wrought-iron frame with either cast or wrought-iron pedestals bolted

to it, and with a pair of bent equalizing levers on each side and a spring between the
wheels, as shown.   The centre plate was carried on a system of bracing, clearly shown
in the engravings.   This form of truck has been built continuously ever since it was
first introduced, with very little change, and has been adopted by other locomotive
builders substantially as it was designed by Mr. Rogers, and probably is more extensively
used, and has given greater satisfaction than any other form of locomotive truck that
has ever been made.   It is still the standard locomotive truck on many railroads.

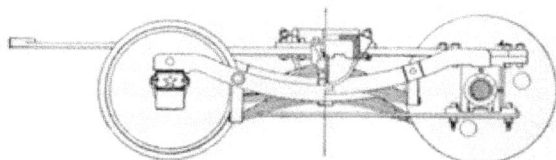

Fig. 207.

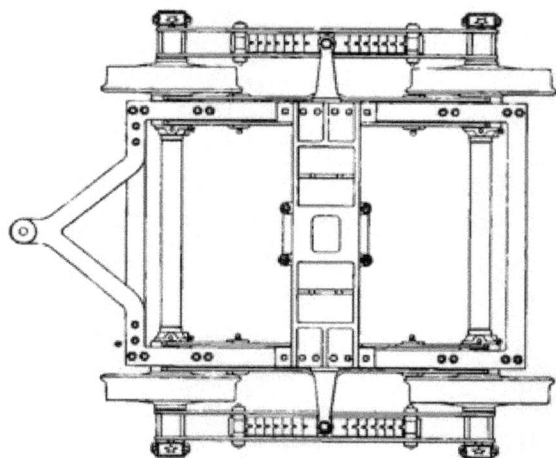

Fig. 208.

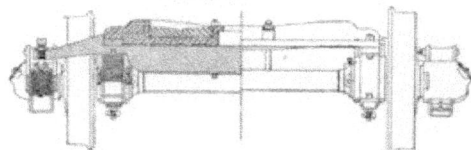

Fig. 209.

Figs. 207, 208, and 209 represent a truck introduced in 1852. This had journal
bearings both inside and outside of the wheels.   It was used for fast passenger engines,

and is shown in Fig. 23. It was first made with a centre bearing, but later the Bissell arrangement, which is shown in the engravings, was combined with it.

In 1857 Mr. Bissell patented the truck, which ever since has been known by his name. His first patent was for a four-wheeled truck, shown by Figs. 210, 211, and 212. The frame of this truck was extended backward, and instead of turning around a centre-pin between the two axles, the pin C, was placed some distance behind the rear axle, and the truck turned or vibrated around it. The engine rested on a pair of V-shaped inclined planes, midway between the two axles. One of these inclined planes is shown in section at D, in Fig. 212.

Fig. 210.

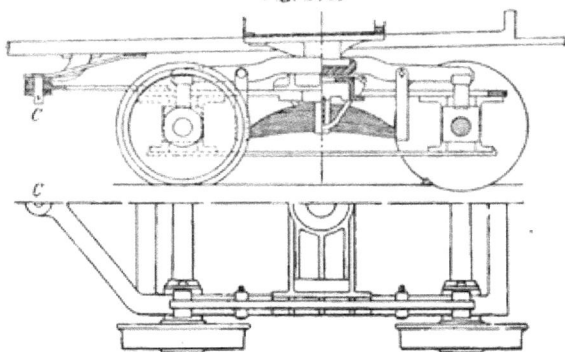

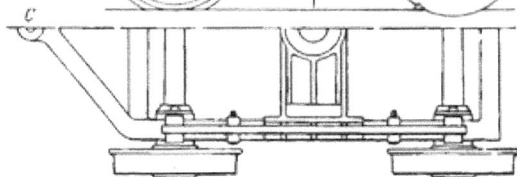

Fig. 211.

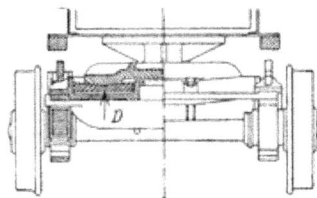

Fig. 212.

The inventor claimed that a truck of his plan adjusts itself to the curvature of the track better than one of the ordinary plan. Mr. Hudson was one of the first to recognize the value of Bissell's invention, and applied it to a locomotive in 1858. In the same year Bissell patented the single axle or "pony" truck, as it is often called. This was constructed on substantially the same principle as his four-wheeled truck, and is represented in Figs.

213, 214, and 215. In the engraving, Fig. 214, to save room, the extension of the frame, which is attached to the centre-pin, is represented as being broken. This truck was applied to some Mogul engines at the Rogers Works in 1863.

Fig. 213.

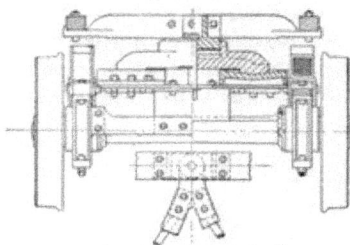

Fig. 215.

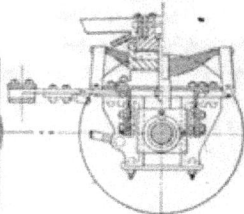

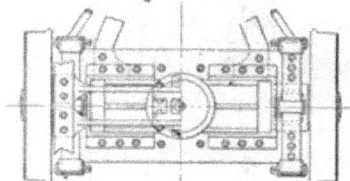

Fig. 214.

In 1862 Mr. Alba F. Smith patented " the employment in a locomotive engine of a truck or pilot fitted with pendent links to allow of lateral motion to the engine."

Fig. 216.

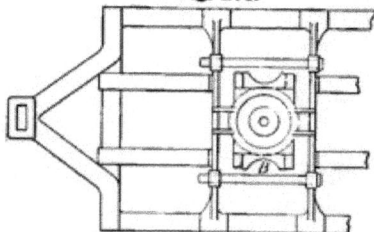

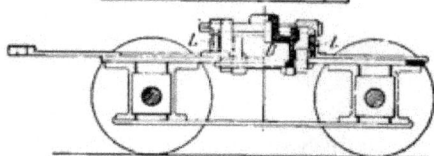

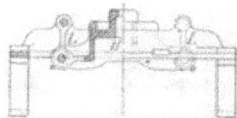

Fig. 217.   **Fig. 218.**

Smith's invention consisted in the substitution of swing links for the inclined planes in Bissell's truck. Smith's truck is shown in Figs. 216, 217, and 218. The engine rested on a bolster B, which was suspended from the truck by swing-links L, L, Figs. 217 and 218. From these it received the name of the swing-motion truck. It was first applied to a locomotive at the Rogers Works in 1865.

In 1864 Mr. Hudson took out another patent for an improvement in lateral moving trucks, which is shown in Figs. 219, 220, and 221. Instead of pivoting the truck to a fixed point behind the axles at A, as Bissell did, Mr. Hudson used a long link or "radius bar" B, B, which was pivoted at its front end C, to a pair of lugs attached to the centre pin plate of the engine. The back end of the radius bar was allowed some lateral motion, but was confined within certain limits by a sort of guide shown at D. This arrangement, Mr. Hudson claimed, permitted the truck to adjust itself more perfectly to curves of different radii than was possible without the use of the radius bar. The arrangement was used with both Bissell's and Smith's lateral motion mechanism.

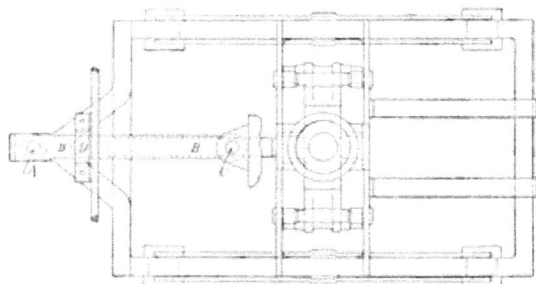

Fig. 219.

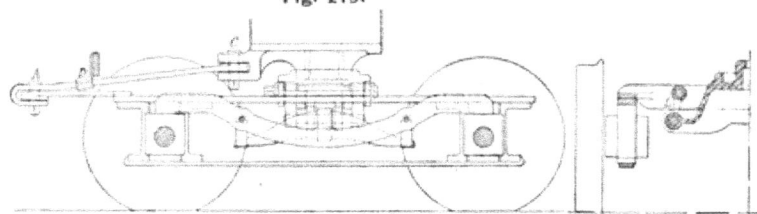

Fig. 220.   Fig. 221.

The most important results accomplished by Bissell's invention were due to the adoption of his lateral moving single axle or "pony truck," as it is called, which was pivoted behind the axle. The first engine of the "Mogul" type, Fig. 26, which was built at the Rogers Works, had a two-wheeled Bissell truck with the inclined planes for producing the lateral motion. It was completed in 1863. Afterwards the swing-links patented by Mr. Smith were used.

In the description of the engine illustrated by Fig. 17, it was pointed out that it had equalizing levers, which extended from the driving axle to the centre of the truck on each side of the engine, with springs in the centre of the levers. Although this arrangement did not give satisfaction at that time, it had the germ of an invention which Mr. Hudson afterwards applied very extensively.

In 1864 he patented the arrangement, shown by Figs. 222, 223, and 224, of an equalizing lever between the two-wheeled truck and the front driving wheels, whereby both the truck and driving wheels maintain their proper portion of weight and accommodate themselves to the vertical, as well as to the lateral motion, required to enable the engine to pass over uneven tracks and around curves with ease as well as with safety. In the arrangement referred to the driving wheels E and F each have the usual springs e and f, connected together by an equalizing lever I, with a fulcrum at i. The front driving wheels G, have springs g. The front strap or hangers n, of these springs are connected to a cross-beam J, (shown clearly in Fig. 223). A central equalizing lever K

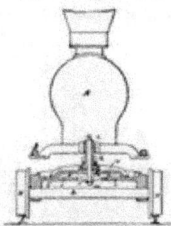

Fig. 222.   Fig. 224.

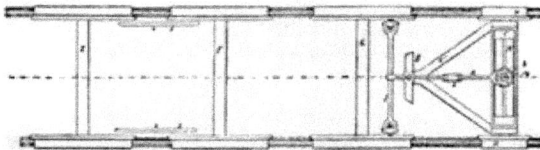

Fig. 223.

bears on the middle of the cross-beam J. It has a fulcrum at k, and its front end rests on the bolster or swing-beam N of the truck. The effect of this arrangement is that any weight borne by the driving wheels is transmitted to the truck and vice versa. In his patent specification Mr. Hudson said:

"If tracks could be made perfectly uniform and regular and be maintained in that condition, my invention would be of little importance; but in practice irregularities more or less serious occur at nearly every joint or junction of the ends of the rails, and at certain points in the track, as in passing switches and across tracks, and especially in passing over small obstacles or defects in the road, the inequality in the load which is thrown upon the several wheels becomes immense; unless, in addition to the use of the springs, provision is made by introducing equalizing levers in some manner, to induce a unity of action between each pair of wheels and some other pair. The three pairs of drivers E, F, and G, Fig. 222, have been connected together by equalizing levers; but I have never known the two pairs E, F, to be connected together into one system, and the forward drivers G, to be connected to the truck wheels, so as to form another and independent system, previous to my invention.

"My invention practically supports the forward portion of the structure at the point k, and the rear portion of the structure on the two points i, i, opposite the sides of the fire-box; thus making a triangle on which the structure is carried with a certainty of holding each wheel with sufficient force upon the track, and yielding easily and safely to every ordinary inequality."

Figs. 222, 223, and 224 are copied from the drawing of the patent specification. In these drawings a truck with Bissell's inclined planes c, c, is represented. Figs. 225, 226, and 227 show the arrangement used by Mr. Hudson in 1865 for Mogul locomotives. In this truck Smith's swing-links were substituted instead of Bissell's inclined planes.

In 1867 Mr. Hudson patented his double-end truck locomotive, to which

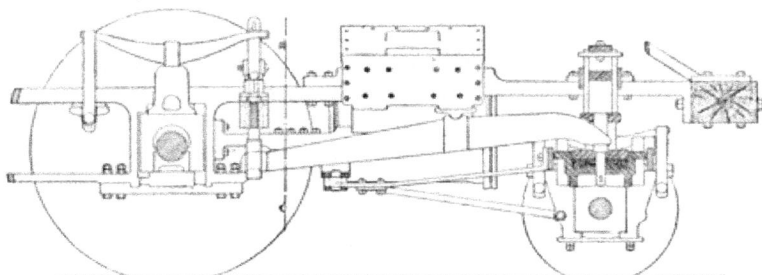

Fig. 225.

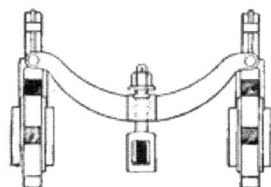

Fig. 226.

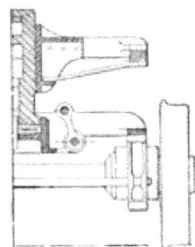

Fig. 227.

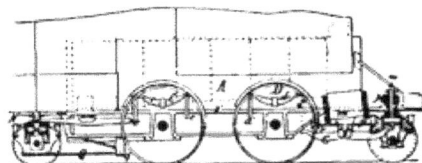

Fig. 228.

Fig. 230.

Fig. 231.

Fig. 229.

reference was made in a previous chapter. Figs. 227, 228, 229, and 230, are copied from his patent specification. In this engine the Bissell truck at each end was connected with the springs of the driving wheels adjoining. The truck of what is ordinarily the

front end of the engine was connected to the driving wheel springs by a single equalizing
lever in the manner already described.   At the opposite end of the engine there were
two equalizing levers, one on each side of the fire-box, as shown in the engraving.

In 1872 Mr. Hudson patented another form of double end truck locomotive,
represented in Plate XVIII.   This had a four-wheeled swing-motion truck behind
the fire-box, and a pony truck in front of the cylinders.   Fig. 232 shows the arrangement of the driving wheel springs and the way that they were connected with the pony
truck by the equalizing levers *E*.   The driving wheel springs were not connected with
the four-wheeled truck.

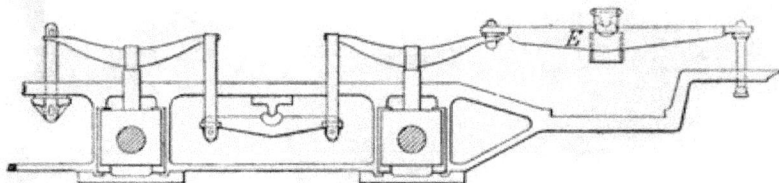

Fig. 232.

As was stated in a previous chapter, in 1872 Mr. Hudson took out a number of
patents covering different forms of truck locomotives to which his method of equalizing
the truck with the driving wheels was applied.   Plates XIX, XXV, XXVI, XXVII,
and XXVIII represent engines built in accordance with some of these plans.

CHAPTER VI.

THE ROGERS LOCOMOTIVE AND MACHINE WORKS IN 1886.

IN 1835 Messrs. Rogers, Ketchum & Grosvenor began some buildings with a view to
the manufacture of locomotives.   The Locomotive Works therefore, are over fifty
years old; although Mr. Rogers and his partners were engaged in machine business some
years before.   During this period not only has the development of the locomotive, considered from an engineering point of view, been very remarkable, but the growth of the
business of their manufacture has been equally so.   The accompanying table shows the
number of locomotives built each year at the Rogers Works up to 1885.   The figures in
the top horizontal line are the diameters of the cylinders of the engines, and the figures
in the vertical columns, under these dimensions, give the number of engines built each

TABLE OF DELIVERY OF LOCOMOTIVE ENGINES

FROM 1837 TO 1885 INCLUSIVE,

Giving Number of Engines and Size of Cylinders of Engines.

YEAR	3½	4½	5	6	7½	8	9	9½	10	10½	11	11½	12	12½	13	13½	14	14½	15	15½	16	16¼	16½	17	17½	18	18½	19	20	Total Engines built
1837											1																			1
1838										1																				7
1839																														11
1840																														7
1841										1																				9
1842										1	2																			6
1843																														9
1844										2	4	2	2																	12
1845		1									2	7	2	1																14
1846										2	4		2	4		3			1	1										17
1847					1	1				3	2	3	2	2	2	3														22
1848					1	2	1	2					1	18		5	2		1					3						39
1849					1					1	3	4	1	5	5	3	2	5		4				8		3				45
1850						3	1	5	4	1	8			5	1	3		7						5						43
1851						1		1	1		8	2	4		11	2	4		14					3						53
1852						1	1	2	8	3	7	3	13	4	18		4		4	2		4	2							68
1853						2	1	12	4	7	21	18	4	3	3	12										4				80
1854						1	1	7	1	6	4	10	36	16	4	12										4				102
1855						2	1	3	2	9	2	5	37	1	41										3					82
1856								1	7	4	8	2	20	45											8					95
1857								7	8	5	3	7	2	8	41											3				84
1858						1		3	1	4	1	4		8		2														24
1859								4	6	4	10	3	26		5															58
1860					2			3	2	4	13	4	7	2	30	2	17	2												88
1861	1			1					1		3	3	2	7		6	3		1											28
1862								2	6	3	5	1	7	2	7	3		2		4										42
1863					2				1		5	5	16	36	5		4													74
1864									3	13	1	30	10	45		2														104
1865						1		5	1	9		20		24	10	6	19													95
1866								2	4	15	1	19	1	31	11	8	11													103
1867	1							1		5	2	14	22	6	1															52
1868								2	1	7	12	1	31	1	1	14														70
1869	1	2					1	10	3	14	19	50	1	11	5															117
1870		1			1			9	3	26	44	56	1	7	4															145
1871			4	3		1	2	2	5	17	38	59	3	19	12															165
1872		1			1	1	1	12	21	70	41	24																		172
1873					1	1	1	10	2	17	1	72	1	55	32															193
1874							1	5	11	2																				19
1875	1			1		1	3	3	1	7	12	13																		42
1876						1	1	2	5	3		5																		17
1877			1			2	2	4	3	2																				14
1878					2	6	1	1	2	4	3	19	3	8																46
1879		1	1		2	1	3	1	3	6	1	31	3	3																56
1880					2		2	5	32	34	9	24	17														125
1881					2		1	1	10	42	37	63	24	41														221
1882							2	18	29	81	85	20	25															260
1883				1			2	5	35	21	84	77	53	1														279
1884				1			2	3	2	6	36	20	3	7														80
1885				12	1		3	1	2	34	20																	73

year, with cylinders of the size indicated above. Although the diameter of the cylinder is not a very exact measure of the capacity of locomotives, nevertheless, in the table it gives a tolerably correct idea of their increase in size since the time Mr. Rogers first started in the business.

The Sandusky, the first locomotive built by him, weighed probably less than ten tons. Since then locomotives have been built at the Rogers Works weighing 57 tons. The figures in the lower right hand corner of the table show that most of the locomotives built in late years had very large cylinders and the engines themselves were of corresponding size.

The engraving of the Works opposite page 3 shows them as they were in 1832, and the frontispiece in 1886. The plan map in Fig. 233 on page 79 will perhaps give a better idea of their magnitude than the perspective view.

The last catalogue of the Rogers Locomotive and Machine Works was issued in 1876. Since then the facilities for doing work has been more than doubled. A large number of tools have been added, some of them specially designed for locomotive building. The shops are all thoroughly equipped with the most approved modern tools for doing accurate work, and with a complete system of templates and gauges, by the use of which the same parts of locomotives are furnished with a degree of precision which insures their being interchangeable. This makes it practicable to supply duplicate parts of locomotives manufactured by the Rogers Locomotive and Machine Works at the shortest notice, which facilitates repairs and reduces very materially the cost of the maintainance of motive power.

About 2,000 men can be advantageously employed, and thirty-three full-sized locomotives can be turned out per month. The late Superintendent, Mr. William S. Hudson, who did so much for the reputation of these Works, has been succeeded by Mr. John Headden, who was formerly with the New Jersey Railroad & Transportation Company before that line was leased to the Pennsylvania Railroad Company.

The Works have, in fact, every facility which long experience, thorough organization and abundant capital can provide for conducting the business of manufacturing locomotives.

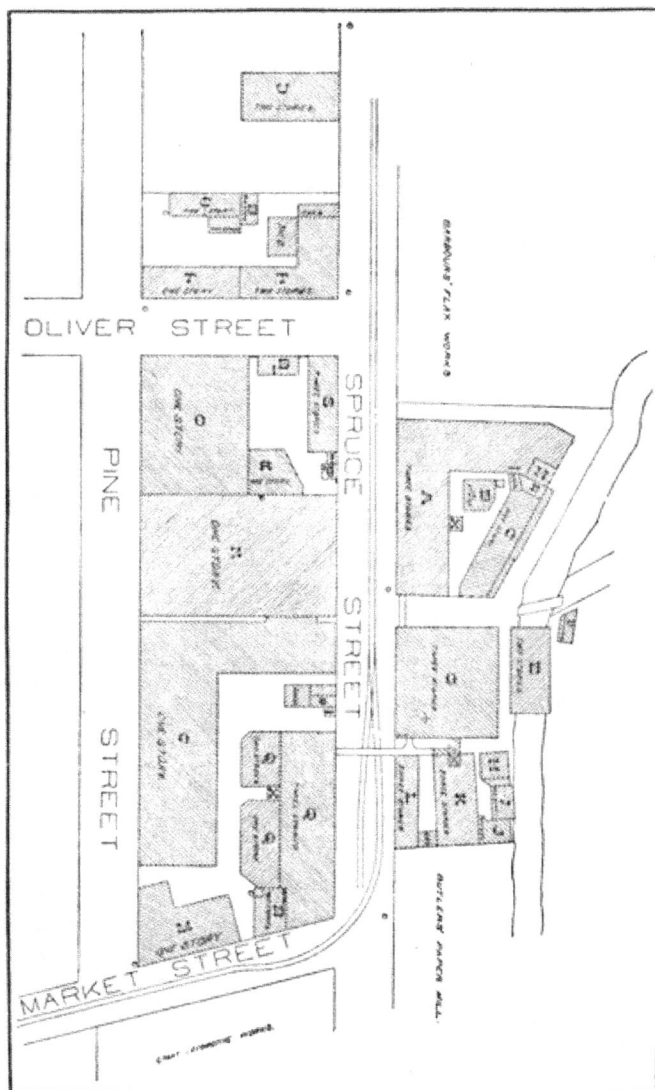

Fig. 233.

CHAPTER VII.

A REMARKABLE RUN OF 426.6 MILES BY ROGERS' LOCOMOTIVES ON THE NEW YORK, WEST SHORE & BUFFALO RAILWAY.

THE following letters to the Editors of the *Railroad Gazette*, which was published in that paper of July 17, 1885, will explain itself:—

NEWARK, N. Y., July 10, 1885.

TO THE EDITOR OF THE RAILROAD GAZETTE:

Herewith I send you a copy of train sheet, showing run made by special train of three cars from East Buffalo to Frankfort yesterday, over the New York West Shore & Buffalo Railway. I wish to call your attention to some of the features of this extraordinary run, to show the perfection of our road-bed and rolling-stock, and the high standard of service which renders it possible to maintain such a high rate of speed without an accident or without delay to other trains. No preparation was made for the train, as we did not know of its coming until a few hours before it left Niagara Falls. Prominent officials of the Baltimore & Ohio, Wabash, Grand Trunk and West Shore Railroads were on board, *en route* for New York. Some of these gentlemen kept an accurate record of the running time and report that several miles were made in 43 seconds, while the greater part of the run was made at a speed averaging 45 to 48 seconds per mile. This is at the rate of 70 to 83 miles per hour. If you will analyze the run you will be surprised to find that their assertions must be true, and that the speed was maintained throughout the whole of the run. Please note the run from East Buffalo to Genesee Junction, 61 miles. Starting from a dead stop at East Buffalo, they came to a stop at Genesee Junction within exactly 56 minutes.

The run from Alabama to Genesee Junction, 36.3 miles, was made in precisely 30 minutes.

The run from East Buffalo to Newark, 93.4 miles, was made in 97 minutes. There are two stoppages to be deducted from this : one of 7 minutes at Genesee Junction for water and oiling engine, and a full stop at Red Creek for the New York, Lake Erie & Western Grade crossing, for which we deduct two minutes — making actual running time 88 minutes.

At Newark the train stopped 9 minutes to change engines.

The conditions were not so favorable for fast running east of Newark as west; but the distance from Newark to Frankfort was covered in 134 minutes ; distance, 108.3 miles. There were six stoppages in this distance, aggregating a delay of 17 minutes, which makes the actual running time 117 minutes.

The whole run from East Buffalo to Frankfort, 202 miles, was made in four hours, or 240 minutes. Deducting total detentions of 35 minutes, the actual running time was 205 minutes.

Between Syracuse and Buffalo we have double track only at intervals, the greater portion being single track. In going in and out of the double-track sections, the train was compelled to run slowly over the Wharton switches. These delays, although not computed, will add something to this very remarkable run.

I submit this as the fastest run ever made in the United States or Canada, and I doubt if it ever has been equalled in the world.

W. H. WHEATLY,

Chief Train Dispatcher.

In order to preserve a permanent and correct record of the remarkable run described by Mr. Wheatly, Mr. Layng, the General Manager of the New York, West Shore & Buffalo line caused elaborate tables to be made out embodying every important fact connected with the performance of their engines, with diagramatic drawings Figs.

234 and 235, showing the principal features of the engines and the table which follows them gives their principal dimensions. A transcript was also made from the schedule

CLASS A (ANTHRACITE)

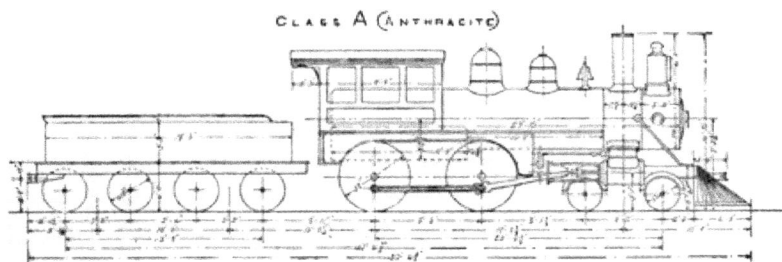

Fig. 234.

CLASS B (BITUMINOUS)

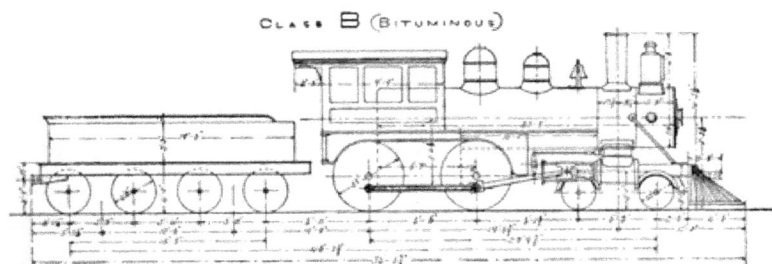

Fig. 235.

board, of which Fig. 236 is a copy, which shows graphically the movement of the train. In this diagram the vertical lines represent time as indicated by the figures in the horizontal line above, and the horizontal lines represent the stations or distance. Their names are given in the column on the right side of the diagram and their distance from Buffalo is given in the column of figures on the left side. The diagonal line through the diagram shows the progress of the train. The inclination of this line indicates its speed.

All the facts relating to this run were given in a large tracing from which blue prints were made and from these the diagrams and the tables which follow Fig. 236 have been reproduced.

LOCOMOTIVES: GENERAL DIMENSIONS, WEIGHTS, ETC.

SPECIFICATIONS.	CLASS A.	CLASS B.
Cylinders, diameter and stroke,	18" × 24"	18" × 24"
" ports-length,	16"	16"
" steam port-width,	1½"	1½"
" exhaust "	3½"	3½"
Valves, Allen Richardson,		
" travel (maximum),	5½"	5½"
" outside lap,	1"	1"
" inside "	1/16"	1/16"
" lead in full gear,	½"	¾"
Exhaust nozzles high double,		
" " diameter,	3½"	3½"
" " height from base of st'k	18"	18"
Boiler pressure per square inch,	140	140
Smoke stack height above rail,	14'–5½"	14'–5½"
" " diameter at top,	20½"	20¾"
" " base,	18"	18"
" " smallest internal diameter,	15"	15"
Engine truck wheels, Allen paper,		
Tender " " "		
Capacity of tender coal (pounds),	15000	15000
" " water (gallons),	3000	3000
Boiler, diameter of smallest ring,	55"	55"
" length of barrel,	10'–11"	10'–11"
Firebox, outside,	10'–9"	6'–7"
" width "	48"	41¾"
" depth inside front,	49¼"	74¾"
" size of grate,	120¾" × 40½"	70¾" × 34½"
" grate area (square feet),	34	17
Tubes, number,	188	188
" ext. diameter,	2"	2"
" length between sheets,	10'–10½"	10'–10½"
" area through (square feet),	3.17	3.17
Heating surface tubes,	1084	1084
" " firebox, "	128	128
" " total "	1212	1212
Weight of engine in working order,	96000	94500
" " " on truck,	32000	32000
" " " drivers,	64000	62500
" " " tender loaded,	76000	76000
" " " empty,	36400	36400
Maximum weight of engine and tender,	172000	170500
Average " " " "	154000	154000

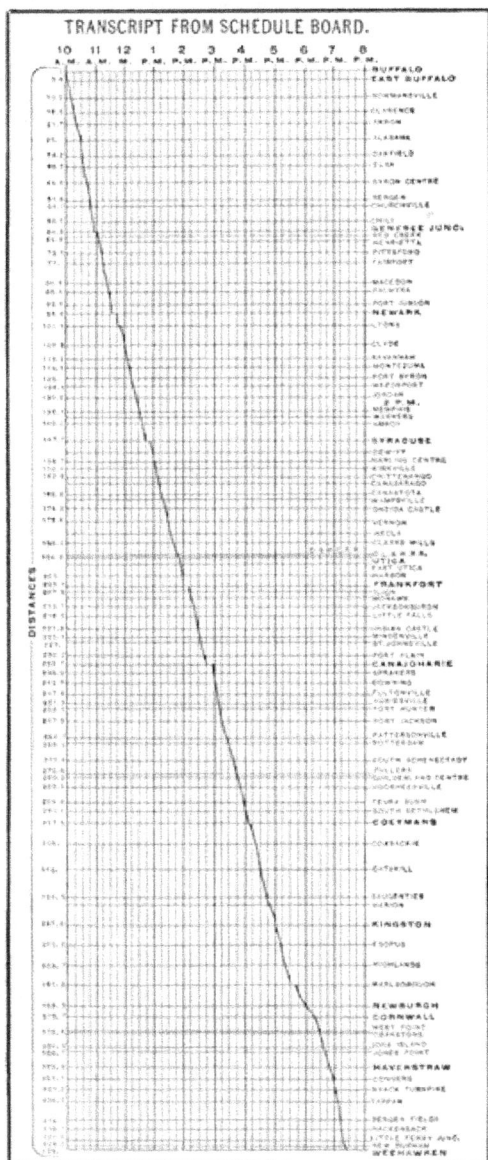

Fig. 236.

RUNNING RECORD: TRANSCRIPT FROM TRAIN DISPATCHER'S SHEETS.

BUFFALO DIVISION.

WEST END.

Dist.	Stations.	Time.	Remarks.
0.0	Buffalo,	10.04	
3.4	East Buffalo,	10.12	Took train from Niagara Falls [Branch.
10.9	Bowmansville,	10.18	
16.5	Clarence,	10.23	
21.7	Akron,	10.30	
28.0	Alabama,	10.35	
34.2	Oakfield,	10.39	
38.5	Elba,	10.44	
44.8	Byron Centre,	10.50	
51.8	Bergen,	10.52	
54.1	Churchville,	11.00	[Took water and oiled.
60.6	Chili,	11.07	
64.3	Genesee Junc.	11.11	Grade crossing B.N.V.&P.R.R.
66.7	Red Creek,	11.11	Grade crossing N.Y.L.E.&W.R.R
72.8	Pittsford,	11.18	
77.0	Fairport,	11.23	
84.6	Macedon,	11.30	
88.4	Palmyra,		
93.2	Port Gibson,	11.37	
96.8	Newark,	11.41	Changed engines.

EAST END.

Dist.	Stations.	Time.	Remarks.
96.8	Newark,	11.50	Held 2 mins. by freight.
101.9	Lyons,	12.06	
109.3	Clyde,	12.12	
115.2	Savannah,	12.16	
118.8	Montezuma,	12.20	
123.0	Port Byron,	12.24	
126.1	Weedsport,	12.30	
130.8	Jordan,	12.35	
136.1	Memphis,	12.40	
140.8	Amboy,	12.48	
147.7	Syracuse,	12.55	Stopped for lunch and took water.
155.7	Manlius Centre	1.04	
158.4	Kirkville,	1.08	
162.3	Chittenango,	1.12	
168.6	Canastota,	1.19	Grade crossing E. C. & N. R. R.
174.2	Oneida Castle,	1.27	
179.4	Vernon,	1.32	
188.0	Clark's Mills,	1.41	Grade crossing D. & H.C.Co.R.R.
194.2	Utica,	1.41	
201.0	Harbor,	1.54	2 grade crossings D. & H. C. D.
205.1	Frankfort,	2.00	[L. & W.
		2.04	Changed engines.

HUDSON RIVER DIVISION.

WEST END.

Dist.	Stations	Time	Remarks
205.1	Frankfort,	2.10	
207.3	Ilion,	2.15	
209.5	Mohawk,	2.17	
213.1	Jacksonburgh,	2.22	
216.9	Little Falls,	2.25	
221.8	Indian Casile,	2.29	
225.1	Mindenville,	2.33	
227.0	St. Johnsville,	2.35	
232.2	Fort Plain,	2.40	
235.7	Canajoharie,	2.48 / 2.56	Stopped for water.
238.9	Sprakers,	3.00	
242.9	Downing,	3.04	
247.6	Fultonville,	3.08	
251.3	Auriesville,	3.11	
253.1	Fort Hunter,	3.13	
257.9	Port Jackson,	3.18	
264.7	Pattersonville,	3.26	Stopped by block.
266.5	Rotterdam Jc.,	3.29	Grade crossing D. & H. C. Co. R. R
273.4	So. Schenectady	3.40	
278.8	Fullers,		
280.2	Guilderland Ct		
283.8	Voorheesville,	3.52	Grade crossing D. & H. C. Co. R. R
289.8	Ffura Bush,	3.54 / 3.58	
293.4	So. Bethlehem,		
297.8	Coeymans,	4.07	Changed engines.

EAST END.

Dist.	Stations	Time	Remarks
297.8	Coeymans,	4.13	
306.0	Coxsackie,	4.24	
316.0	Catskill,	4.32	
326.9	Saugerties,	4.45	Stopped 3 minutes for water at [Glenerie.
337.8	Kingston,	5.01 / 5.04	Grade crossing U. & D. R. R.
345.6	Esopus,	5.12	
353.7	Highland,	5.21	
361.3	Marlborough,	5.30	
369.3	Newburgh,	5.39 / 5.59 / 6.00 / 6.20	Stopped by block.　"　　"　　"
373.7	Cornwall,	6.20	
378.6	West Point,	6.27	
384.9	Iona Island,	6.37½	
386.7	Jones Point,	6.40	
392.8	Haverstraw,	6.50	
397.5	Congers,	6.58	Stopped by block.
401.8	Nyack Turnpike	7.02	
406.9	Tappan,	7.07	
414.0	Bergenfields,	7.13	
418.1	Hackensack,	7.16	
421.6	Little Ferry J,	7.19	
424.5	New Durham,	7.23	Stopped by block.
426.0	Weehawken,	7.27	

SPEED BETWEEN STATIONS,

IN NEAREST EVEN MILES PER HOUR,

DEDUCTING ACTUAL STOPS PLUS 2 MINUTES PER STOP FOR LOSING AND GAINING HEADWAY.

BUFFALO DIVISION.

EAST END.

Station columns (left to right): Newark, Lyons, Clyde, Savannah, Montezuma, Port Byron, Weedsport, Jordan, Memphis, Amboy, Syracuse, Manlius Centre, Kirkville, Chittenango, Canastota, Oneida Castle, Vernon, Clarks Mills, Utica, Harbor, Frankfort.

Row stations: Newark, Lyons, Clyde, Savannah, Montezuma, Port Byron, Weedsport, Jordan, Memphis, Amboy, Syracuse, Manlius Centre, Kirkville, Chittenango, Canastota, Oneida Castle, Vernon, Clarks Mills, Utica, Harbor, Frankfort.

WEST END.

Station columns (left to right): East Buffalo, Bowmansville, Clarence, Akron, Alabama, Oakfield, Elba, Byron Centre, Bergen, Churchville, Chili, Genesee Junc., Red Creek, Pittsford, Fairport, Macedon, Palmyra, Port Gibson, Newark.

Row stations: East Buffalo, Bowmansville, Clarence, Akron, Alabama, Oakfield, Elba, Byron Centre, Bergen, Churchville, Chili, Genesee Junc., Red Creek, Pittsford, Fairport, Macedon, Palmyra, Port Gibson, Newark.

HUDSON RIVER DIVISION.

EAST END.

Column stations (mileage/distance table):
Weehawken, New Durham, Little Ferry Jc, Hackensack, Bergenfields, Tappan, Nyack Turnpike, Congers, Haverstraw, Jones Point, Iona Island, West Point, Cornwall, Newburgh, Marlborough, Highland, Esopus, Kingston, Saugerties, Catskill, Coxsackie, Coeymans.

Row stations:
Coeymans, Coxsackie, Catskill, Saugerties, Kingston, Esopus, Highland, Marlborough, Newburgh, Cornwall, West Point, Iona Island, Jones Point, Haverstraw, Congers, Nyack Turnpike, Tappan, Bergenfields, Hackensack, Little Ferry Jc, New Durham, Weehawken.

WEST END.

Column stations:
Coeymans, So. Bethlehem, Feura Bush, Voorheesville, Guilderland Ctr, Fullers, So. Schenectady, Rotterdam Junc, Pattersonville, Port Jackson, Fort Hunter, Auriesville, Fultonville, Downing, Speakers, Canajoharie, Fort Plain, St. Johnsville, Mindenville, Indian Castle, Little Falls, Jacksonburgh, Mohawk, Ilion, Frankfort.

Row stations:
Frankfort, Ilion, Mohawk, Jacksonburgh, Little Falls, Indian Castle, Mindenville, St. Johnsville, Fort Plain, Canajoharie, Sprakers, Downing, Fultonville, Auriesville, Fort Hunter, Port Jackson, Pattersonville, Rotterdam Jc, So. Schenectady, Fullers, Guilderland Ctr, Voorheesville, Feura Bush, So. Bethlehem, Coeymans.

MAKE UP AND WEIGHT OF TRAIN.

BUFFALO DIVISION.

WEST END.

ENGINEMAN CHAS. SMITH, CONDUCTOR FULLER.

ENGINE,	No. 45, Class B, (Bituminous)	94500.
TENDER,	With ⅔ load of coal and water,	62800.
BAGGAGE CAR,	West Shore, No. 737	46030.
OFFICIAL "	" 90	61200.
" "	Balto. & Ohio " 711	46430.
	TOTAL POUNDS,	310,960.
	TOTAL TONS,	155 +

EAST END.

ENGINEMAN MARTIN PIERCE, CONDUCTOR FULLER.

ENGINE,	No. 50, Class B, (Bituminous)	94500.
TENDER,	With ⅔ load of coal and water,	62800.
BAGGAGE CAR,	West Shore, No. 737	46030.
OFFICIAL "	" 90	61200.
" "	Balto. & Ohio " 711	46430.
	TOTAL POUNDS,	310,960.
	TOTAL TONS,	155 +

HUDSON RIVER DIVISION.

WEST END.

ENGINEMAN JERRY LYNCH, CONDUCTOR SHULTZ.

ENGINE,	No. 27, Class A, (Anthracite)	96000.
TENDER,	With ⅔ load of coal and water,	62800.
BAGGAGE CAR,	West Shore, No. 737	46030.
OFFICIAL "	" 90	61200.
" "	Balto. & Ohio " 711	46430.
	TOTAL POUNDS,	312,460.
	TOTAL TONS,	156 +

EAST END.

ENGINEMAN JOHN JONES, CONDUCTOR SHULTZ.

ENGINE,	No. 36, Class B, (Bituminous)	94500.
TENDER,	With ⅔ load of coal and water,	62800.
BAGGAGE CAR,	West Shore, No. 737	46030.
OFFICIAL "	" 90	61200.
" "	Balto. & Ohio " 711	46430.
	TOTAL POUNDS,	310,960.
	TOTAL TONS,	155 +

SUMMARIZED RUNNING RECORD.

ALL DIVISIONS.

PRINCIPAL STATIONS ONLY.

DIST.	STATIONS.	TIME.	REMARKS.
0.0	BUFFALO,	——	
3.4	EAST BUFFALO,	10.04	Took train from Niag. Falls branch.
64.3	GENESEE JC.,	11.00 11.07	B. N. Y. & P. Grade crossing. Took water and oiled.
96.8	NEWARK,	11.41 11.50	Changed engines.
147.7	SYRACUSE,	12.48 12.55	Stopped for lunch and took water.
194.2	UTICA,	1.54	
205.1	FRANKFORT,	2.04 2.10	Changed engines.
235.7	CANAJOHARIE,	2.48 2.56	Stopped for water.
266.5	ROTTERDAM J.,	3.29 3.29	Stopped by block.
297.8	COEYMANS,	4.07 4.13	Changed engines.
337.8	KINGSTON,	5.01	
373.7	CORNWALL,	6.20 6.20	
392.8	HAVERSTRAW,	6.50 6.51	Stopped by block.
426.0	WEEHAWKEN,	7.27	

SUMMARIZED SPEED RECORD.

WITH DEDUCTIONS AS NOTED.

ALL DIVISIONS.

PRINCIPAL STATIONS ONLY.

	BUFFALO.	EAST BUFFALO.	GENESEE JUNC.	NEWARK.	SYRACUSE.	UTICA.	FRANKFORT.	CANAJOHARIE.	ROTTERDAM, J.	COEYMANS.	KINGSTON.	CORNWALL,	HAVERSTRAW.	WEEHAWKEN.
BUFFALO,	—	—	—	—	—	—	—	—	—	—	—	—	—	—
EAST BUFFALO		—	68	67	64	62	62	60	60	60	58	58	57	57
GENESEE JUN.			—	65	61	59	60	58	58	59	56	56	55	56
NEWARK,				—	59	57	59	57	57	58	55	55	54	55
SYRACUSE,					—	56	58	56	57	57	55	54	54	55
UTICA,						—	73	55	57	58	54	54	53	54
FRANKFORT,							—	51	55	57	53	53	52	54
CANAJOHARIE,								—	60	60	54	54	53	54
ROTTERDAM, J.									—	61	52	51	51	53
COEYMANS,										—	46	50	49	52
KINGSTON,											—	54	51	55
CORNWALL,												—	46	55
HAVERSTRAW,													—	62
WEEHAWKEN,														—

PHYSICAL CHARACTERISTICS OF ROAD.

ELEMENTS.		BUFF. DIV.		HUD. RIV. DIV.		ENTIRE LINE.
		W. END.	E. END.	W. END.	E. END.	
PROFILE.						
Level,		16.93	27.03	29.94	42.18	30.18
Up grade (going east)	Percentages.	45.59	37.50	29.59	30.41	35.62
Down " " "		37.48	35.47	40.47	27.41	34.20
Up grade " "	Average feet per mile.	16.25	15.88	16.36	16.96	16.32
Down grade " "		18.75	18.49	17.66	22.77	19.51
ALIGNMENT.						
Tangents,	Percentages.	82.52	83.44	73.85	84.26	82.06
Curves.		17.48	16.56	26.15	15.74	17.94
Curvature : average : degrees :		1°–20'–57"	1°–46'–11"	2°–4'–40"	1°–48'–17"	1°–46'–34"

Four different locomotives were used: One of these, **number 45, ran** from Buffalo to Newark, 96.8 miles; another, number 50, took the train **from** Newark to Frankfort, 108.3 miles; the third, engine number 27, was used from Frankfort **to** Coeymans, 9.27, and the last part of the journey from Coeymans to Weehawken, 128.2 miles was made with engine number 36. Engine number 27 was an anthracite coal burner, Fig. 234 with a long fire-box of the type known as Class A, on the West Shore road. Nos. 45, 50, and 36 were bituminous coal burners, Fig. 235 designated Class B engines. They were all built at the Rogers Locomotive Works in Paterson, N. J., from the design of the late Howard Fry.

Another remarkable run was made by a Rogers engine **on the New** York, West Shore & Buffalo Railway on the 8th of October, 1885, the particulars of which are given in the following table from which it will be seen that this train at times **attained** the remarkable speed of 80 miles per hour. The table was prepared for Mr. J. D. Layng, the general manager of the line:

Memorandum of speed made on Special, consisting of Engine No. 43, John Davis, Engineer, with Car No. 100, October 8, 1885, on run between Genesee Junction and East Buffalo.

Miles.	Seconds.	Miles per H.	Miles.	Seconds.	Miles per H.
1	52	69	1	52	69
1	55	62	1	51	71
1	50	72	1	49	73
1	51	71	1	47	77
1	53	68	1	50	72
1	49	73	1	45	80
1	50	72	1	45	80
1	50	72	1	45	80
1	49	73	1	50	72
1	53	68	1	48	75
			1	52	69
10	512	70	11	534	74

CHAPTER VIII.

THE TRACTIVE POWER OF LOCOMOTIVES.*

IT may be stated, generally, that a locomotive exerts its power in drawing trains by means of the friction or adhesion of the driving wheels on the rails. Or, to quote from Pambour's old "Treatise on Locomotive Engines:" "Two conditions are necessary in order that an engine may draw a given load: First, that the dimensions and proportions of the engine and its boiler enable it to produce on the piston, by means of the steam, the necessary pressure, which constitutes what is properly termed the power of the engine; and second, that the weight of the engine be such as to give a sufficient adhesion to the wheel on the rail. These two conditions of power and weight must be in concordance with each other; for, if there is a great power of steam and little adhesion, the latter will limit the effect of the engine, and there will be steam lost; if, on the other hand, there is too much weight for the steam, that weight will be a useless burden, the limit of the load being in that case marked by the steam."

There is a good deal of difference in the figures given by various authorities to indicate the proportion which the friction or adhesion of the wheels on the rails bears to the weight on them. The figures which are perhaps used most in practice are those published in Molesworth's "Pocket-Book of Engineering Formulæ." These are as follows:

ADHESION PER TON OF 2,240 LBS. ON THE DRIVING-WHEELS.

When the rails are very dry, ...600 lbs. per ton
When the rails are very wet,...550 " " "
In ordinary English weather,...450 " " "
In misty weather, if the rails are greasy, ...300 " " "
In frosty or snowy weather, ...200 " " "

In D. K. Clark's "Manual for Mechanical Engineers, page 724, he gives a report of experiments made by M. Poirée on the Paris & Lyons Railroad with a wagon by skidding the wheels. Of these experiments Clark says:

"At speeds under 20 miles per hour it appears from the table that, when the rails are dry, the co-efficient of friction, or the adhesion, is one-fifth of the weight, and that on very dry rails it is one-fourth. As the speed is increased, the adhesion is reduced. These data are corroborative of the results of the author's experiments on the ultimate tractive force of locomotives on dry rails, from which he obtained a co-efficient of friction equal to one-fifth of the weight, at speeds of about 10 miles per hour."

In the paper "On the Effect of Brakes upon Railway Trains" read by Captain Galton before the Institution of Mechanical Engineers,† the following determination of

* A considerable portion of this chapter is reprinted from the *Railroad Gazette* of June 6, 1879.
† See *Engineering* of May 2, 1879, Page 371.

the adhesion of wheels is given. It must be kept in mind, too, that he makes the distinction between "adhesive" and sliding friction. By "adhesive" is meant the friction between rolling wheels and the track:

"On dry rails it was found that the co-efficient of adhesion of the wheels was generally over 0.20. In some cases it rose to 0.25 or even higher. On wet or greasy rails without sand, it fell as low as 0.15 in an experiment, but averaged about 0.18. With the use of sand on wet rails it was above 0.20 at all times; and when the sand was applied at the moment of starting, so that the wind of the rotating wheels did not blow it away, it rose up to 0.35, and even above 0.40."

This is probably the most correct determination of the adhesion of wheels that has ever been made, and shows that the ordinary rule of taking the adhesion at *one-fifth* of the weight in the driving wheels is quite within the limits of ordinary practice. Even on a wet or greasy rail, with the use of sand, it was above 0.20 at all times. In fact, if we want to calculate the maximum power which a locomotive will exert if the rails are sanded, we might take the adhesion at *one-third*, and under favorable conditions without sand would be quite safe at *one-fourth*.

In order to put these figures in a form in which they can easily be remembered and conveniently used, they may be given as follows:

ADHESION OF LOCOMOTIVES.

Under ordinary conditions without using sand on the rails, or on wet sanded rails...............	One-fifth the weight on the driving-wheels.
Under favorable conditions without sand, ...	One-fourth the weight on the driving-wheels.
On a dry, sanded rail, ...	One-third the weight on the driving-wheels.

These may be taken as working data, but before the tractive power of a locomotive can be determined it must be known how much power is required to draw a given load over a road with known grades and curves. If the authorities be consulted with reference to this point a wider difference even than that relating to the adhesion of driving-wheels will be found to exist. Without comparing these, it may be stated that the most recent experiments have shown that the resistance of good American cars does not exceed 6 lbs. per ton of 2,000 lbs. at very slow speeds on a straight and level track, and when in the best condition and good weather it is probably not over 4 lbs. The wind, however, has an important influence, and as this is very variable it is hardly safe to take the resistance, under the conditions named above, at less than 6 lbs. per ton.

With reference to the influence of speed on the resistance, it must be admitted that our knowledge is very inexact, and probably the law or laws which govern it are not understood. The following rule, though, will give results which do not differ materially from those given by the most reliable experiments which have thus far been made.

To get the resistance per ton (of 2,000 lbs.) of a train on a straight and level track at any given speed :

Square the speed in miles per hour and divide by 171 and add 6.

To get the resistance per ton due to any grade :

Multiply the rise in feet per mile by 0.3788 and add the quotient to the resistance due to the speed on a straight and level track.

Our knowledge of the resistance due to curves, like that due to speed, is in a very unsatisfactory condition, but the most reliable information we have indicates that the resistance is equal to about half a pound per ton per degree of curvature.

We may then tabulate these calculations as follows :

RESISTANCE OF TRAINS.

On straight and level track at very low speeds,................	6 lbs. per ton of 2,000 lbs.		
For resistance due to speed : *Square the speed in miles per hour and divide by 171,*......	— ''	''	''
For resistance due to grade : *Multiply the rise in feet per mile by 0.3788,*	— ''	''	''
For resistance due to curves : *Add ½ lb. per degree of curvature,*........	— ''	''	''
Total,..	— lbs.	''	''

If the radius of the curve is given, the "degree" may be found approximately *by dividing the radius into 5730.* This rule is correct enough for ordinary curves of over 500 feet radius.

Having these data, suppose we want to calculate how much, say, a Consolidation engine will pull up a grade of 70 feet per mile, with 9° curves and at a speed of 20 miles per hour. The first question to determine will be whether we want to know the maximum load which such an engine will draw, or what it will do in good weather, or what it will do at all times, excepting in snow storms. In the first case we would take the adhesion at ⅓ the weight on the driving-wheels; in the second at ¼, and in the last case at ⅕. We will assume that the second represents our hypothetical case, and that the locomotive has a weight of 11,000 lbs. on each driving-wheel, or a total of 88,000 lbs. The adhesion would therefore be one-fourth of 88,000 lbs. = 22,000 lbs. The train resistance per ton would be as follows :

$$\text{Resistance on straight and level track} = 6.0 \text{ lbs.}$$

$$'' \quad \text{due to speed} = \frac{20 \times 20}{171} = 2.3 ''$$

$$'' \quad '' \quad '' \text{ grade} = 70 \times 0.3788 = 26.5 ''$$

$$'' \quad '' \quad '' \text{ curve} = 9 \times \tfrac{1}{2} = 4.5 ''$$

$$\text{Total,}............... .... 39.3 \text{ lbs.}$$

Therefore, as each ton will have a resistance of 39.3 lbs., and as our engine is capable of exerting a tractive force of 22,000 lbs., the total load which it can pull would be represented by

$$\frac{22,000}{39.3} = 559.8 \text{ tons.}$$

As the engine and tender weigh about 72 tons, the train which our engine will pull will be represented by 559.8 — 72 = 487.8. Of course, to do this work the cylinders must be large enough to turn the wheels, and the boiler have the requisite capacity to supply steam. It is very rare that a locomotive has not cylinder capacity sufficient to turn the wheels. It happens much oftener that the cylinders of Locomotives are *too large* instead of too small. This is due to the fact that the boiler pressure has of late years been much increased while the size of the cylinders has not been diminished in the same proportion.

The table on the following page, which gives the resistance in lbs. per ton (of 2,000 lbs.) is taken from Forney's Catechism of the Locomotive, and was calculated by the rules given above. The various speeds are indicated in the headings at the tops of the columns, and the rate of gradients, that is the rise in feet per mile is given in the first column on the left. The resistance on a grade of say 50 feet per mile, and a speed of 20 miles per hour can be found on the horizontal line opposite the figure 50 in the first column and in the column under the heading of that speed, and is 27.2 lbs. per ton.

The simplest way to calculate how heavy a train a locomotive will pull under ordinary conditions of weather is to *divide the weight on all the driving wheels by 4, which will give their adhesion to the rails. Then divide this by the resistance for the required grade and speed, taken from the table. If curves and grades occur simultaneously, add to the resistance given in the table ½ lb. for each degree of the curve.* The quotient will be the weight of the train *including* that of the tender and locomotive, which the latter will pull on the grade and at the speed given.

The capacity of the Rogers engines, in the tables on the following pages, is calculated for an adhesion equal to one-fourth of the weight on the driving wheels. The maximum capacity of these engines, under very favorable conditions, will be somewhat greater than that given in the tables.

TABLE OF RESISTANCES OF RAILROAD TRAINS,

ON A STRAIGHT TRACK,

WITH DIFFERENT GRADES AND SPEEDS.

Rise of gradient, feet per mile.	Resistance due to ascent alone in pounds per ton (2000 lbs.) of train.	Total resistance, pounds per ton, at rate of 5 miles per hour.	10 miles per hour.	15 miles per hour.	20 miles per hour.	25 miles per hour.	30 miles per hour.	35 miles per hour.	40 miles per hour.	45 miles per hour.	50 miles per hour.	60 miles per hour.	70 miles per hour.
0		6.1	6.6	7.3	8.3	9.6	11.2	13.1	15.3	17.8	20.6	27.0	34.6
5	1.8	7.9	8.4	9.1	10.1	11.4	13.0	14.9	17.1	19.6	22.4	28.8	36.4
10	3.7	9.8	10.3	11.0	12.0	13.4	14.9	16.8	19.0	21.5	24.3	30.7	38.3
15	5.6	11.7	12.2	12.9	13.9	15.2	16.8	18.7	21.9	24.4	27.2	33.6	41.2
20	7.5	13.6	14.1	14.8	15.8	17.1	18.7	20.6	22.8	25.3	28.1	34.5	42.1
25	9.4	15.5	16.0	16.7	17.7	19.0	20.6	22.5	24.7	27.2	31.0	37.4	45.0
30	11.3	17.4	17.9	18.6	19.6	21.9	22.5	24.4	26.6	29.1	31.9	38.3	45.9
35	13.2	19.3	19.8	20.5	21.5	22.8	24.4	26.3	28.5	31.0	33.8	40.2	47.8
40	15.1	21.2	21.7	22.4	23.4	24.7	26.3	28.2	30.4	32.9	35.7	42.1	49.7
45	17.0	23.1	23.6	24.3	25.3	26.6	28.2	30.1	32.3	34.8	37.6	44.0	51.6
50	18.9	25.0	25.5	26.2	27.2	28.5	30.1	32.0	34.2	36.7	39.5	45.9	53.5
60	22.7	28.8	29.3	30.0	31.0	32.3	33.9	35.8	38.0	40.5	43.5	49.9	57.5
70	26.5	32.6	33.1	33.8	34.8	36.1	37.7	39.6	41.8	44.3	47.1	53.5	61.1
80	30.3	36.4	36.9	37.6	38.6	39.9	40.5	42.4	44.6	47.1	49.9	56.3	63.9
90	34.0	41.0	40.6	41.3	42.3	43.6	45.2	47.1	49.3	51.8	54.6	61.0	68.6
100	37.8	43.9	44.4	45.1	46.1	47.4	49.0	51.9	54.1	56.6	59.4	65.8	73.4
110	41.6	47.7	48.2	48.9	49.9	51.2	52.8	54.7	56.9	59.4	62.2	68.6	76.2
120	45.4	51.5	52.0	52.7	53.7	55.0	56.6	58.5	60.7	63.2	66.0	72.4	80.0
130	49.2	55.3	55.8	56.5	57.5	58.8	60.4	62.3	64.5	67.0	69.8	76.2	83.8
140	53.0	59.1	59.6	60.3	61.3	62.6	64.2	66.1	68.3	70.8	73.6	80.0	87.6
150	56.8	62.9	63.4	64.1	65.1	66.4	68.0	69.9	72.1	74.6	77.4	83.8	91.4
160	60.6	66.7	67.2	67.9	68.9	70.2	71.8	73.7	75.9	78.4	81.2	87.6	95.2
170	64.3	70.4	70.9	71.6	72.6	73.9	75.5	77.4	79.6	82.1	84.9	91.3	98.9
180	68.1	74.2	74.7	75.4	76.4	77.7	79.3	81.2	83.4	85.9	88.7	95.1	102.7
190	71.9	78.0	78.5	79.2	80.2	81.5	83.1	85.0	87.2	89.7	92.5	98.9	106.5
200	75.7	81.8	82.3	83.0	84.0	85.3	86.9	88.8	91.0	93.5	96.3	102.7	110.3
210	79.5	85.6	86.1	86.8	87.8	89.1	90.7	92.6	94.8	97.3	100.1	106.5	114.1
220	83.3	89.4	89.9	90.6	91.6	92.9	94.5	96.4	98.6	101.1	103.9	110.3	117.6
230	87.1	93.2	93.7	94.4	95.4	96.7	98.3	100.2	102.4	104.9	107.7	114.1	121.7
240	90.8	96.9	97.4	98.1	99.1	100.4	102.0	103.9	106.1	108.6	111.4	117.8	125.4
250	94.6	100.7	101.2	101.9	102.9	103.2	105.8	107.7	109.9	112.4	115.2	121.6	129.2
260	98.4	104.5	105.0	105.7	106.7	107.0	108.6	110.5	112.7	115.2	118.0	124.4	132.0
270	102.2	108.3	108.8	109.5	110.5	111.8	113.4	115.3	117.5	120.0	122.8	129.2	136.8
280	106.0	112.1	112.6	113.3	114.3	115.6	117.2	119.1	121.3	123.8	126.6	133.0	140.6
290	109.8	115.9	116.4	117.1	118.1	119.4	121.0	122.9	125.1	127.6	130.4	136.8	144.4
300	113.6	119.7	120.2	120.9	121.9	123.2	124.8	126.7	128.9	131.4	134.2	140.6	148.2

PLATES

— AND —

TABLES OF DIMENSIONS AND CAPACITY

— OF —

LOCOMOTIVES.

CHAPTER IX.

PLATES AND TABLES OF DIMENSIONS AND CAPACITY OF LOCOMOTIVES OF
4 FT., 8½ IN. GAUGE OR WIDER.

IN the following tables the principal dimensions, weight, etc., and the calculated capacity for hauling loads is given for the different classes of locomotives manufactured by the Rogers Locomotive and Machine Works. In making the calculations the adhesion of the engines as mentioned in the preceding chapter, was taken at *one-fourth* the weight on the driving wheels. Experience has shown that the adhesion of the driving wheels is fully equal to that proportion of the weight on them, in good weather and under favorable conditions. The calculations are made for straight lines and for the grades and speeds specified in each table. An allowance, which has been explained in the previous chapter, must be made for curves.

PLATE I.

Eight Wheel Standard Locomotives

FOR PASSENGERS.

Gauge 4 ft., 8½ in. or wider.   Fuel, Bituminous Coal.

General Design shown by Plate I.

	Cylinders. Diameter and Stroke. inches.	Dia'eter of Driving Wheels. inches.	Wheel Base.		Weight, in running order. Pounds.			Separate Tender. Capacity of Tank. Gals.
			Of Driving Wheels.	Total.	On Driving Wheels. lbs.	On Truck. lbs.	Total. lbs.	
1	15 × 22	66	7 ft. 9 in.	21 ft. 4 in.	41200	23200	64400	1800
2	15 × 24	66	7 ft. 9 in.	21 ft. 8 in.	42200	23400	65600	2000
3	16 × 22	66	7 ft. 9 in.	21 ft. 7 in.	43200	23400	66600	2000
4	16 × 24	66	8 ft.	22 ft.	44200	24400	68600	2200
5	17 × 22	66	8 ft.	21 ft. 9½ in.	45200	24400	69600	2200

	Load in tons of 2000 pounds in addition to Engine and Tender, at 30 miles an hour, on a grade per mile of								
	On a Level.	10 ft.	20 ft.	40 ft.	60 ft.	80 ft.	100 ft.	125 ft.	150 ft.
1	869	640	500	341	253	203	159	125	100
2	888	654	510	347	257	206	161	126	101
3	910	671	523	357	264	213	166	130	105
4	940	685	534	363	269	216	168	131	105
5	950	700	546	372	275	221	173	135	108

PLATE II.

Eight Wheel Standard Locomotives

FOR PASSENGERS.

Gauge, 4 ft., 8½ in. or wider.　Fuel, Bituminous Coal.

General Design shown by Plate II.

	Cylinders. Diameter and Stroke. inches.	Dia'eter of Driving Wheels. inches.	Wheel Base.		Weight, in running order. POUNDS.			Separate Tender.
			Of Driving Wheels.	Total.	On Driving Wheels. lbs.	On Truck. lbs.	Total. lbs.	Capacity of Tank. Gals.
1	17 × 24	66	8 ft. 6 in.	22 ft. 8½ in.	48700	27000	75700	2300
2	18 × 22	66	8 ft. 6 in.	22 ft. 6½ in.	49200	27200	76400	2300
3	18 × 24	66	8 ft. 6 in.	22 f. 11½ in	52400	27600	80000	2600
4	19 × 22	66	8 ft. 6 in.	22 ft. 7½ in.	53000	28000	81000	2600
5								

	Load in tons of 2000 pounds in addition to Engine and Tender, at 30 miles an hour, on a grade per mile of								
	On a Level.	10 ft.	20 ft.	40 ft.	60 ft.	80 ft.	100 ft.	125 ft.	150 ft.
1	1024	754	588	400	296	237	185	145	116
2	1035	762	595	405	300	241	188	147	118
3	1101	811	632	430	318	256	199	155	125
4	1114	820	640	435	322	258	201	157	126
5									

PLATE III.

Eight Wheel Standard Locomotives

FOR PASSENGERS OR FREIGHT.

Gauge 4 ft., 8½ in. or wider.   Fuel, Bituminous Coal.

General Design shown by Plate III.

	Cylinders. Diameter and Stroke. inches.	Dia'eter of Driving Wheels. inches.	Wheel Base.		Weight, in running order. POUNDS			Separate Tender.
			Of Driving Wheels.	Total.	On Driving Wheels. lbs.	On Truck. lbs.	Total. lbs.	Capacity of Tank. Gals.
1	15 × 24	56	7 ft. 9 in.	21 ft. 7 in.	41000	23000	64000	2000
2	16 × 22	56	7 ft. 9 in.	21 ft. 6 in.	42000	23000	65000	2000
3	16 × 24	56	8 ft.	21 ft. 11 in.	43000	24000	67000	2200
4	17 × 22	56	8 ft.	21 ft. 9½ in.	44000	24000	68000	2200
5	17 × 24	56	8 ft. 3 in.	22 ft. 4½ in.	47500	26500	74000	2300

Load in tons of 2000 pounds in addition to Engine and Tender, at 20 miles an hour, on a grade per mile of								
On a Level.	10 ft.	20 ft.	40 ft.	60 ft.	80 ft.	100 ft.	125 ft.	150 ft.
1	1184	803	598	387	280	214	171	133	106
2	1212	822	611	396	286	219	175	135	108
3	1239	840	624	403	291	222	177	137	109
4	1268	861	639	413	298	228	182	140	112
5	1368	927	689	445	321	246	195	151	120

Eight Wheel Standard Locomotives

FOR PASSENGERS OR FREIGHT.

Gauge, 4 ft., 8½ in. or wider. Fuel, Bituminous Coal.

General Design shown by Plate III.

	Cylinders. Diameter and Stroke. inches.	Dia'eter of Driving Wheels. inches.	Wheel Base.		Weight, in running order. POUNDS.			Separate Tender. Capacity of Tank. Gals.
			Of Driving Wheels.	Total.	On Driving Wheels. lbs.	On Truck. lbs.	Total. lbs.	
1	15 × 22	62	7 ft. 9 in.	21 ft. 3 in.	40700	23000	63700	1800
2	15 × 24	62	7 ft. 9 in.	21 ft. 7 in.	41700	23200	64900	2000
3	16 × 22	62	7 ft. 9 in.	21 ft. 6 in.	42700	23200	65900	2000
4	16 × 24	62	8 ft.	21 ft. 11 in.	43700	24200	67900	2200
5	17 × 22	62	8 ft.	21 ft. 9½ in.	44700	24200	68900	2200

	Load in tons of 2000 pounds in addition to Engine and Tender, at 20 miles an hour, on a grade per mile of								
	On a Level.	10 ft.	20 ft.	40 ft.	60 ft.	80 ft.	100 ft.	125 ft.	150 ft.
1	1175	797	593	384	277	213	170	132	105
2	1203	816	607	392	283	217	173	134	107
3	1232	835	622	402	290	222	177	138	110
4	1259	853	634	410	295	226	180	139	111
5	1289	873	649	420	302	231	184	143	114

Eight Wheel Standard Locomotives

FOR PASSENGERS OR FREIGHT.

Gauge 4 ft., 8 ½ in. or wider.   Fuel, Bituminous Coal.

General Design shown by Plate III.

	Cylinders. Diameter and Stroke. inches.	Dia'eter of Driving Wheels. inches.	Wheel Base.		Weight, in running order. POUNDS.			Separate Tender.
			Of Driving Wheels.	Total.	On Driving Wheels. lbs.	On Truck. lbs.	Total. lbs.	Capacity of Tank. Gals.
1	17 × 24	62	8 ft. 3 in.	22 ft. 4 ½ in.	48200	26800	75000	2300
2	18 × 22	62	8 ft. 3 in.	22 ft. 3 ½ in.	48500	27000	75500	2300
3								
4								
5								

	Load in tons of 2000 pounds in addition to Engine and Tender, at 20 miles an hour, on a grade per mile of								
	On a Level.	10 ft.	20 ft.	40 ft.	60 ft.	80 ft.	100 ft.	125 ft.	150 ft.
1	1389	942	701	453	327	249	199	154	123
2	1397	947	704	455	328	251	200	155	123
3									
4									
5									

PLATE IV.

Eight Wheel Standard Locomotives

FOR PASSENGERS.

Gauge, 4 ft., 8½ in. or wider.   Fuel, Anthracite Coal.

General Design shown by Plate IV.

	Cylinders. Diameter and Stroke. inches.	Dia'eter of Driving Wheels. inches.	Wheel Base.		Weight, in running order. POUNDS.			Separate Tender.
			Of Driving Wheels.	Total.	On Driving Wheels. lbs.	On Truck. lbs.	Total. lbs.	Capacity of Tank. Gals.
1	17 × 22	66	8 ft.	21 ft. 11 in.	59500	23500	83000	2200
2	17 × 24	66	8 ft.	22 ft. 1½ in.	63000	25000	88000	2300
3	18 × 22	66	8 ft.	21 ft. 11 in.	63000	25000	88000	2300
4	18 × 24	66	8 ft. 6 in.	22 ft. 9½ in.	64000	32500	96500	2600
5								

	Load in tons of 2000 pounds in addition to Engine and Tender, at 30 miles an hour, on a grade per mile of							
	On a Level.	10 ft.	20 ft.	40 ft.	60 ft.	80 ft.	100 ft.	125 ft.	150 ft.
1	1264	934	731	502	375	303	239	190	155
2	1337	988	773	530	396	320	252	200	163
3	1337	988	773	530	396	320	252	200	163
4	1352	997	780	532	396	319	250	197	159
5									

PLATE V.

Eight Wheel Standard Locomotives

FOR PASSENGERS OR FREIGHT.

Gauge, 4 ft., 8½ in. or wider.   Fuel, Anthracite Coal.

General Design shown by Plate V.

	Cylinders. Diameter and Stroke. inches.	Dia'eter of Driving Wheels. inches.	Wheel Base.		Weight, in running order. POUNDS.			Separate Tender.
			Of Driving Wheels.	Total.	On Driving Wheels. lbs.	On Truck. lbs.	Total. lbs.	Capacity of Tank, Gals.
1	16 × 24	62	8 ft.	22 ft. 1½ in.	59500	23500	83000	2200
2								
3								
4								
5								

	Load in tons of 2000 pounds in addition to Engine and Tender, at 30 miles an hour, on a grade per mile of							
	On a Level.	10 ft.	20 ft.	40 ft.	60 ft.	80 ft.	100 ft.	125 ft.	150 ft.
1	1264	934	731	501	375	303	239	190	155
2									
3									
4									
5									

PLATE VI.

Mogul Locomotives

FOR FREIGHT.

Gauge 4 ft., 8½ in. or wider.   **Fuel, Bituminous Coal.**

General Design shown by Plate VI.

	Cylinders. Diameter and Stroke. inches.	Dia'eter of Driving Wheels. inches.	Wheel Base.		Weight, in running order. POUNDS.			Separate Tender.
			Of Driving Wheels.	Total.	On Driving Wheels. lbs.	On Truck. lbs.	Total. lbs.	Capacity of Tank. Gals.
1	16 × 24	48	14 ft. 10 in	22 ft. 4 in.	57000	13000	70000	2200
2	17 × 24	48	15 ft.	22 ft. 6 in.	60500	13500	74000	2300
3								
4								
5								

	On a Level.	10 ft.	20 ft.	40 ft.	60 ft.	80 ft.	100 ft.	125 ft.	150 ft.
	Load in tons of 2000 pounds in addition to Engine and Tender, at 20 miles an hour, on a grade per mile of								
1	1659	1129	844	549	401	311	251	197	160
2	1760	1198	895	584	426	330	266	210	170
3									
4									
5									

Mogul Locomotives

FOR FREIGHT.

Gauge, 4 ft., 8½ in. or wider. Fuel, Bituminous Coal.

General Design shown by Plate VI.

	Cylinders. Diameter and Stroke. inches.	Dia'eter of Driving Wheels. inches.	Wheel Base.		Weight, in running order. POUNDS.			Separate Tender.
			Of Driving Wheels.	Total.	On Driving Wheels. lbs.	On Truck. lbs.	Total. lbs.	Capacity of Tank. Gals.
1	18 × 24	50	15 ft. 2 in.	22 ft. 8 in.	66500	14500	81000	2600
2								
3								
4								
5								

	Load in tons of 2000 pounds in addition to Engine and Tender, at 20 miles an hour, on a grade per mile of								
	On a Level.	10 ft.	20 ft.	40 ft.	60 ft.	80 ft.	100 ft.	125 ft.	150 ft.
1	1935	1317	984	642	468	363	293	230	187
2									
3									
4									
5									

Mogul Locomotives

FOR FREIGHT

Gauge, 4 ft.. 8 ½ in. or wider.   **Fuel**, Bituminous Coal.

General Design shown by Plate VI.

	Cylinders. Diameter and Stroke. inches.	Dia'eter of Driving Wheels. inches.	Wheel Base.		Weight, in running order. POUNDS.			Separate Tender.
			Of Driving Wheels.	Total.	On Driving Wheels. lbs.	On Truck. lbs.	Total. lbs.	Capacity of Tank. Gals.
1	18 × 24	54	15 ft. 2 in.	22 ft. 9 in.	67800	14200	82000	2600
2	19 × 22	54	12 ft. 10 in	20 ft. 5 ½ in.	68000	14400	82400	2600
3	19 × 24	54	15 ft. 4 in.	23 ft.	70000	16000	86000	3000
4	20 × 24	54	15 ft. 6 in.	23 ft. 2 in.	78400	19950	98350	3000
5								

	Load in tons of 2000 pounds in addition to Engine and Tender, at 20 miles an hour, on a grade per mile of								
	On a Level.	10 ft.	20 ft.	40 ft.	60 ft.	80 ft.	100 ft.	125 ft.	150 ft.
1	1973	1343	1003	655	478	370	299	235	191
2	1979	1347	1006	657	480	371	300	236	192
3	2034	1384	1033	674	490	379	306	240	195
4	2281	1553	1160	758	552	428	345	272	221
5									

PLATE VII.

Mogul Locomotives

FOR FREIGHT.

Gauge, 4 ft., 8½ in. or wider.   Fuel, Anthracite Coal.

General Design shown by Plate VII.

	Cylinders. Diameter and Stroke. inches.	Dia'eter of Driving Wheels. inches.	Wheel Base.		Weight, in running order. POUNDS.			Separate Tender.
			Of Driving Wheels.	Total.	On Driving Wheels. lbs.	On Truck. lbs.	Total. lbs.	Capacity of Tank. Gals.
1	18×24	48	13 ft. 6 in.	20 ft. 10 in.	73000	12000	85000	2600
2	19×24	48	13 ft. 6 in.	20 ft. 10 in.	78000	13000	91000	2600
3								
4								
5								

Load in tons of 2000 pounds in addition to Engine and Tender, at 20 miles an hour, on a grade per mile of

	On a Level.	10 ft.	20 ft.	40 ft.	60 ft.	80 ft.	100 ft.	125 ft.	150 ft.
1	2127	1449	1084	709	518	402	325	257	209
2	2275	1551	1160	759	555	431	349	276	226
3									
4									
5									

Mogul Locomotives

FOR FREIGHT.

Gauge, 4 ft., 8½ in. or wider.   Fuel, Anthracite Coal.

General Design shown by Plate VII.

	Cylinders. Diameter and Stroke. inches.	Dia'eter of Driving Wheels. inches.	Wheel Base.		Weight, in running order. POUNDS.			Separate Tender. Capacity of Tank. Gals.
			Of Driving Wheels.	Total.	On Driving Wheels. lbs.	On Truck. lbs.	Total. lbs.	
1	19 × 24	54	13 ft. 6 in.	20 ft. 10 in.	79500	13500	93000	3000
2	20 × 24	54	14 ft	21 ft. 6½ in.	85500	14500	100000	3000
3								
4								
5								

	Load in tons of 2000 pounds in addition to Engine and Tender, at 20 miles an hour, on a grade per mile of								
	On a Level.	10 ft.	20 ft.	40 ft.	60 ft.	80 ft.	100 ft.	125 ft.	150 ft.
1	2316	1578	1179	771	563	437	353	279	227
2	2494	1696	1268	832	608	473	383	303	247
3									
4									
5									

PLATE VIII.

Ten Wheel Locomotives

FOR FREIGHT.

Gauge, 4 ft., 8½ in. or wider.   Fuel, Bituminous Coal.

General Design shown by Plate VIII.

	Cylinders. Diameter and Stroke. inches.	Dia'eter of Driving Wheels. inches.	Wheel Base.		Weight, in running order. POUNDS.			Separate Tender.
			Of Driving Wheels.	Total.	On Driving Wheels. lbs.	On Truck. lbs.	Total. lbs.	Capacity of Tank. Gals.
1	16 × 24	50	12 ft. 6 in.	22 ft. 9 in.	54000	18500	72500	2200
2	17 × 22	50	12 ft. 6 in.	22 ft. 8 in.	54000	19000	73000	2200
3	17 × 24	50	13 ft. 3 in.	23 ft. 5 in.	56500	21000	77500	2300
4	18 × 22	50	13 ft. 3 in.	23 ft. 4 in.	57000	21000	78000	2300
5								

	Load in tons of 2000 pounds in addition to Engine and Tender, at 20 miles an hour, on a grade per mile of								
	On a Level.	10 ft.	20 ft.	40 ft.	60 ft.	80 ft.	100 ft.	125 ft.	150 ft.
1	1567	1066	795	518	376	291	234	183	148
2	1582	1076	803	523	380	294	236	186	150
3	1637	1113	830	540	393	302	242	189	153
4	1652	1123	838	543	396	305	245	192	155
5									

Ten Wheel Locomotives

FOR FREIGHT.

Gauge, 4 ft., 8½ in. or wider.   Fuel, Bituminous Coal.

General Design shown by Plate VIII.

	Cylinders. Diameter and Stroke. inches.	Dia'eter of Driving Wheels. inches.	Wheel Base.		Weight, in running order. POUNDS.			Separate Tender. Capacity of Tank. Gals.
			Of Driving Wheels.	Total.	On Driving Wheels. lbs.	On Truck. lbs.	Total. lbs.	
1	16×24	54	12 ft. 6 in.	22 ft. 9 in.	55000	18500	73500	2200
2	17×22	54	12 ft. 6 in.	22 ft. 8 in.	54800	19200	74000	2200
3	17×24	54	13 ft. 3 in.	23 ft. 5 in.	57000	21500	78500	2300
4	18×22	54	13 ft. 3 in.	23 ft. 4 in.	57500	21500	79000	2300
5	18×24	54	13 ft. 4 in.	23 ft. 6 in.	61500	23500	85000	2600

Load in tons of 2000 pounds in addition to Engine and Tender, at 20 miles an hour, on a grade per mile of

	On a Level.	10 ft.	20 ft.	40 ft.	60 ft.	80 ft.	100 ft.	125 ft.	150 ft.
1	1596	1085	810	528	383	296	238	187	151
2	1590	1081	807	525	382	295	237	186	150
3	1652	1123	838	545	396	305	245	192	155
4	1667	1133	846	550	400	308	248	194	157
5	1782	1211	903	587	426	328	263	206	166

Ten Wheel Locomotives

FOR FREIGHT.

Gauge, 4 ft., 8½ in. or wider.   Fuel, Bituminous Coal.

General Design shown by Plate VIII.

	Cylinders. Diameter and Stroke. inches.	Dia'eter of Driving Wheels. inches.	Wheel Base.		Weight, in running order. POUNDS.			Separate Tender.
			Of Driving Wheels.	Total.	On Driving Wheels. lbs.	On Truck. lbs.	Total. lbs.	Capacity of Tank. Gals.
1	19 × 24	54	13 ft. 6 in.	23 ft. 8 in.	65000	25000	90000	3000
2								
3								
4								
5								

	On a Level.	10 ft.	20 ft.	40 ft.	60 ft.	80 ft.	100 ft.	125 ft.	150 ft.
Load in tons of 2000 pounds in addition to Engine and Tender, at 20 miles an hour, on a grade per mile of									
1	1881	1278	952	618	448	345	276	216	174
2									
3									
4									
5									

PLATE IX.

Ten Wheel Locomotives

FOR FREIGHT.

Gauge, 4 ft., 8½ in. or wider.   Fuel, Wood.

General Design shown by Plate IX.

	Cylinders. Diameter and Stroke. inches.	Dia'eter of Driving Wheels. inches.	Wheel Base.		Weight, in running order. POUNDS.			Separate Tender.
			Of Driving Wheels.	Total.	On Driving Wheels. lbs.	On Truck. lbs.	Total. lbs.	Capacity of Tank. Gals.
1	16 × 24	56	12 ft. 6 in.	22 ft. 9 in.	55500	18500	74000	2200
2	17 × 22	56	12 ft. 6 in.	22 ft. 8 in.	55700	19300	75000	2200
3	17 × 24	56	13 ft. 3 in.	23 ft. 5 in.	58000	21500	79500	2300
4	18 × 22	56	13 ft. 3 in.	23 ft. 4 in.	58500	21500	80000	2300
5	18 × 24	56	13 ft. 4 in.	23 ft. 6 in.	62500	23500	86000	2600

Load in tons of 2000 pounds in addition to Engine and Tender, at 20 miles an hour, on a grade per mile of

	On a Level.	10 ft.	20 ft.	40 ft.	60 ft.	80 ft.	100 ft.	125 ft.	150 ft.
1	1611	1096	818	533	387	299	241	189	153
2	1616	1099	820	534	388	300	241	189	153
3	1682	1143	853	555	403	311	249	195	159
4	1697	1153	861	560	407	314	252	197	160
5	1811	1231	922	597	433	333	268	209	169

Ten Wheel Locomotives

FOR FREIGHT.

Gauge, 4 ft., 8½ in. or wider.   Fuel, Wood.

General Design shown by Plate IX.

	Cylinders. Diameter and Stroke. inches.	Dia'eter of Driving Wheels. inches.	Wheel Base.		Weight, in running order. POUNDS.			Separate Tender.
			Of Driving Wheels.	Total.	On Driving Wheels. lbs.	On Truck. lbs.	Total. lbs.	Capacity of Tank. Gals.
1	19×22	56	13 ft. 4 in.	23 ft. 6 in.	63500	24500	88000	2600
2	19×24	56	13 ft. 6 in.	23 ft. 8 in.	68800	25200	94000	3000
3								
4								
5								

	Load in tons of 2000 pounds in addition to Engine and Tender, at 20 miles an hour, on a grade per mile of								
	On a Level.	10 ft.	20 ft.	40 ft.	60 ft.	80 ft.	100 ft.	125 ft.	150 ft.
1	1840	1250	932	606	440	339	272	213	171
2	1994	1355	1010	657	477	367	295	231	186
3									
4									
5									

PLATE X.

Ten Wheel Locomotives

FOR FREIGHT.

Gauge, 4 ft., 8½ in. or wider.   Fuel, Bituminous Coal.

General Design shown by Plate X.

	Cylinders. Diameter and Stroke. inches.	Dia'eter of Driving Wheels. inches.	Wheel Base.		Weight, in running order. POUNDS.			Separate Tender. Capacity of Tank. Gals.
			Of Driving Wheels.	Total.	On Driving Wheels. lbs.	On Truck. lbs.	Total. lbs.	
1	19 × 24	50	15 ft. 1 in.	26 ft.	76000	16000	92000	2600
2								
3								
4								
5								

	On a Level.	10 ft.	20 ft.	40 ft.	60 ft.	80 ft.	100 ft.	125 ft.	150 ft.
	Load in tons of 2000 pounds in addition to Engine and Tender, at 20 miles an hour, on a grade per mile of								
1	2215	1500	1128	738	539	418	338	267	218
2									
3									
4									
5									

Ten Wheel Locomotives

FOR FREIGHT.

Gauge, 4 ft., 8½ in. or wider.   Fuel, Bituminous Coal.

General Design shown by Plate X.

	Cylinders. Diameter and Stroke. inches.	Dia'eter of Driving Wheels. inches.	Wheel Base.		Weight, in running order. POUNDS.			Separate Tender.
			Of Driving Wheels.	Total.	On Driving Wheels. lbs.	On Truck. lbs.	Total. lbs.	Capacity of Tank. Gals.
1	20×24	54	15 ft. 5 in.	26 ft. 3½ in.	80000	18000	98000	3000
2								
3								
4								
5								

	On a Level.	10 ft.	20 ft.	40 ft.	60 ft.	80 ft.	100 ft.	125 ft.	150 ft.
Load in tons of 2000 pounds in addition to Engine and Tender, at 20 miles an hour, on a grade per mile of									
1	2329	1586	1185	775	565	438	354	279	227
2									
3									
4									
5									

.

PLATE XI.

Consolidation Locomotive

FOR FREIGHT.

Gauge, 4 ft., 8½ in. or wider.   Fuel, Bituminous Coal.

General Design shown by Plate XI.

	Cylinders. Diameter and Stroke. inches.	Dia'eter of Driving Wheels. inches.	Wheel Base.		Weight, in running order. POUNDS.			Separate Tender.
			Of Driving Wheels.	Total.	On Driving Wheels. lbs.	On Truck. lbs.	Total. lbs.	Capacity of Tank. Gals.
1	20 × 24	50	14 ft. 9 in.	22 ft. 10 in.	86500	13500	100000	3000
2								
3								
4								
5								

	Load in tons of 2000 pounds in addition to Engine and Tender, at 15 miles an hour, on a grade per mile of								
	On a Level.	10 ft.	20 ft.	40 ft.	60 ft.	80 ft.	100 ft.	125 ft.	150 ft.
1	2881	1884	1380	884	640	494	398	314	256
2									
3									
4									
5									

PLATE XII.

Consolidation Locomotive

FOR FREIGHT.

Gauge, 4 ft., 8½ in. or wider.   Fuel, Anthracite Coal.

General Design shown by Plate XII.

	Cylinders. Diameter and Stroke. inches.	Dia'eter of Driving Wheels. inches.	Wheel Base.		Weight, in running order. POUNDS.			Separate Tender.
			Of Driving Wheels.	Total.	On Driving Wheels. lbs.	On Truck. lbs.	Total. lbs.	Capacity of Tank. Gals.
1	20×24	50	14 ft. 9 in.	22 ft. 10 in.	92800	13600	106400	3000
2								
3								
4								
5								

	Load in tons of 2000 pounds in addition to Engine and Tender, at 15 miles an hour, on a grade per mile of								
	On a Level.	10 ft.	20 ft.	40 ft.	60 ft.	80 ft.	100 ft.	125 ft.	150 ft.
1	3094	2025	1483	951	689	533	430	340	278
2									
3									
4									
5									

PLATE XIII.

Four Wheel Locomotives

FOR SWITCHING.

Gauge, 4 ft., 8½ in. or wider. **Fuel, Bituminous Coal.**

General Design shown by Plate XIII.

	Cylinders. Diameter and Stroke. inches.	Dia'eter of Driving Wheels. inches.	Wheel Base.		Weight, in running order. POUNDS.			Separate Tender.
			Of Driving Wheels.	Total.	On Driving Wheels. lbs.		Total. lbs.	Capacity of Tank. Gals.
1	12 × 20	46	7 ft.	7 ft.	41500		41500	1200
2	12 × 22	46	7 ft.	7 ft.	43500		43500	1200
3	13 × 22	46	7 ft.	7 ft.	45500		45500	1200
4								
5								

Load in tons of 2000 pounds in addition to Engine and Tender, at 10 miles an hour, on a grade per mile of

	On a Level.	10 ft.	20 ft	40 ft.	60 ft.	80 ft.	100 ft.	125 ft.	150 ft.
1	1554	983	709	448	322	248	201	159	130
2	1629	1031	743	470	338	261	211	167	136
3	1704	1078	778	492	354	274	221	175	143
4									
5									

Four Wheel Locomotives

FOR SWITCHING.

Gauge, 4 ft., 8½ in. or wider.   Fuel, Bituminous Coal.

General Design shown by Plate XIII.

| | Cylinders. Diameter and Stroke. inches. | Dia'eter of Driving Wheels. inches. | Wheel Base. | | Weight, in running order. POUNDS. | | Separate Tender. |
			Of Driving Wheels.	Total.	On Driving Wheels. lbs.	Total. lbs.	Capacity of Tank. Gals.
1	12×20	50	7 ft.	7 ft.	43000	43000	1200
2	12×22	50	7 ft.	7 ft.	44000	44000	1200
3	13×22	50	7 ft.	7 ft.	46000	46000	1200
4	14×22	50	7 ft. 6 in.	7 ft. 6 in.	48000	48000	1500
5	15×22	50	7 ft. 6 in.	7 ft. 6 in.	51000	51000	1500

Load in tons of 2000 pounds in addition to Engine and Tender, at 10 miles an hour, on a grade per mile of

	On a Level.	10 ft.	20 ft.	40 ft.	60 ft.	80 ft.	100 ft.	125 ft.	150 ft.
1	1591	1006	725	458	330	254	205	162	133
2	1629	1030	743	470	338	261	211	167	136
3	1704	1078	778	492	354	274	221	175	143
4	1777	1124	810	512	368	284	229	182	148
5	1888	1194	861	545	392	303	244	194	150

Four Wheel Locomotives

FOR SWITCHING.

Gauge, 4 ft., 8½ in. or wider.   Fuel, Bituminous Coal.

General Design shown by Plate XIII.

	Cylinders. Diameter and Stroke. inches.	Dia'eter of Driving Wheels. inches.	Wheel Base.		Weight, in running order. POUNDS.			Separate Tender.
			Of Driving Wheels.	Total.	On Driving Wheels. lbs.		Total. lbs.	Capacity of Tank. Gals.
1	15 × 24	50	7 ft. 6 in.	7 ft. 6 in.	54000		54000	1800
2	16 × 22	50	7 ft. 6 in.	7 ft. 6 in.	54000		54000	1800
3	16 × 24	50	7 ft. 6 in.	7 ft. 6 in.	55500		55500	1800
4								
5								

Load in tons of 2000 pounds in addition to Engine and Tender, at 10 miles an hour, on a grade per mile of

	On a Level.	10 ft.	20 ft.	40 ft.	60 ft.	80 ft.	100 ft.	125 ft.	150 ft.
1	1999	1264	911	576	415	320	258	204	167
2	1999	1264	911	576	415	320	258	204	167
3	2074	1312	946	598	431	332	268	213	174
4									
5									

Four Wheel Locomotives

FOR SWITCHING.

Gauge, 4 ft., 8½ in. or wider. Fuel, Bituminous Coal.

*General Design shown by Plate **XIII**.*

	Cylinders. Diameter and Stroke. inches.	Dia'eter of Driving Wheels. inches.	Wheel Base.		Weight, in running order. POUNDS.			Separate Tender.
			Of Driving Wheels.	Total.	On Driving Wheels. lbs.		Total. lbs.	Capacity of Tank. Gals.
1	15 × 24	54	7 ft. 6 in.	7 ft. 6 in.	54500		54500	1800
2	16 × 22	54	7 ft. 6 in.	7 ft. 6 in.	54500		54500	1800
3	16 × 24	54	7 ft. 6 in.	7 ft. 6 in.	56000		56000	2000
4								
5								

	Load in tons of 2000 pounds in addition to Engine and Tender, at 10 miles an hour, on a grade per mile of								
	On a Level.	10 ft.	20 ft.	40 ft.	60 ft.	80 ft.	100 ft.	125 ft.	150 ft.
1	2036	1287	928	587	422	326	263	208	170
2	2036	1287	928	587	422	326	263	208	170
3	2072	1310	944	596	429	330	266	211	172
4									
5									

PLATE XIV.

Six Wheel Locomotives

FOR SWITCHING.

Gauge, 4 ft., 8½ in. or wider.   Fuel, Bituminous Coal.

General Design shown by Plate XIV.

	Cylinders, Diameter and Stroke. inches.	Dia'eter of Driving Wheels, inches.	Wheel Base.		Weight, in running order. POUNDS.			Separate Tender.
			Of Driving Wheels.	Total.	On Driving Wheels. lbs.		Total. lbs.	Capacity of Tank. Gals.
1	11 × 16	31	10 ft. 3 in.	10 ft. 3 in.	39000		39000	1200
2								
3								
4								
5								

Load in tons of 2000 pounds in addition to Engine and Tender, at 10 miles an hour, on a grade per mile of

	On a Level.	10 ft.	20 ft.	40 ft.	60 ft.	80 ft.	100 ft.	125 ft.	150 ft.
1	1443	913	657	415	299	230	185	147	119
2									
3									
4									
5									

Six Wheel Locomotives

FOR SWITCHING.

Gauge, 4 ft., 8 ½ in. or wider.   Fuel, Bituminous Coal.

General Design shown by Plate XIV.

	Cylinders. Diameter and Stroke. inches.	Dia'eter of Driving Wheels. inches.	Wheel Base.		Weight, in running order. POUNDS.			Separate Tender.
			Of Driving Wheels.	Total.	On Driving Wheels. lbs.		Total. lbs.	Capacity of Tank. Gals.
1	11 × 16	44	10 ft. 3 in.	10 ft. 3 in.	40000		40000	1200
2	14 × 22	44	10 ft. 3 in.	10 ft. 3 in.	61500		61500	1500
3	15 × 22	44	10 ft. 3 in.	10 ft. 3 in.	64500		64500	1500
4	16 × 20	44	10 ft. 3 in.	10 ft. 3 in.	65000		65000	1500
5								

	Load in tons of 2000 pounds in addition to Engine and Tender, at 10 miles an hour, on a grade per mile of								
	On a Level.	10 ft.	20 ft.	40 ft.	60 ft.	80 ft.	100 ft.	125 ft.	150 ft.
1	1480	936	674	426	306	236	191	151	123
2	2300	1460	1052	666	481	372	301	236	196
3	2375	1504	1085	688	497	383	311	248	205
4	2412	1528	1102	699	505	390	316	251	207
5									

Six Wheel Locomotives

FOR SWITCHING.

Gauge, 4 ft., 8½ in. or wider.   Fuel, Bituminous Coal.

General Design shown by Plate XIV.

	Cylinders. Diameter and Stroke. inches.	Dia'eter of Driving Wheels. inches.	Wheel Base.		Weight, in running order. POUNDS.			Separate Tender.
			Of Driving Wheels.	Total.	On Driving Wheels. lbs.		Total. lbs.	Capacity of Tank. Gals.
1	14 × 22	46	10 ft. 3 in.	10 ft. 3 in.	63000		63000	1500
2	15 × 22	46	10 ft. 3 in.	10 ft. 3 in.	65500		65500	1500
3	16 × 20	46	10 ft. 3 in.	10 ft. 3 in.	66000		66000	1500
4	16 × 22	46	10 ft. 3 in.	10 ft. 3 in.	68000		68000	1800
5	16 × 24	46	10 ft. 7 in.	10 ft. 7 in.	72000		72000	1800

	Load in tons of 2000 pounds in addition to Engine and Tender, at 10 miles an hour, on a grade per mile of								
	On a Level.	10 ft.	20 ft.	40 ft.	60 ft.	80 ft.	100 ft.	125 ft.	150 ft.
1	2338	1481	1069	678	489	379	307	244	200
2	2450	1551	1120	710	513	397	322	256	210
3	2450	1551	1120	710	513	397	322	256	210
4	2523	1597	1153	730	527	408	330	262	215
5	2672	1692	1222	774	559	433	350	279	229

Six Wheel Locomotives

FOR SWITCHING.

Gauge, 4 ft., 8 ½ in. or wider.   Fuel, Bituminous Coal.

General Design shown by Plate XIV.

	Cylinders. Diameter and Stroke. inches.	Dia'eter of Driving Wheels. inches.	Wheel Base. Of Driving Wheels.	Wheel Base. Total.	Weight, in running order. POUNDS. On Driving Wheels. lbs.	Weight, in running order. POUNDS.	Weight, in running order. POUNDS. Total. lbs.	Separate Tender. Capacity of Tank. Gals.
1	17 × 22	46	10 ft. 3 in.	10 ft. 3 in.	72500		72500	1800
2	17 × 24	46	10 ft. 7 in.	10 ft. 7 in.	78000		78000	1800
3	18 × 22	46	10 ft. 3 in.	10 ft. 3 in.	78800		78800	2000
4	18 × 24	46	10 ft. 7 in.	10 ft. 7 in.	84000		84000	2000
5	19 × 22	46	10 ft. 3 in.	10 ft. 3 in.	85000		85000	2300

	Load in tons of 2000 pounds in addition to Engine and Tender, at 10 miles an hour, on a grade per mile of								
	On a Level.	10 ft.	20 ft.	40 ft.	60 ft.	80 ft.	100 ft.	125 ft.	150 ft.
1	2672	1692	1222	774	559	433	350	279	229
2	2896	1835	1324	841	607	470	381	304	249
3	2932	1856	1339	849	613	474	384	305	261
4	3118	1975	1426	905	654	506	411	326	268
5	3152	1996	1440	912	658	509	412	327	268

Six Wheel Locomotives

FOR SWITCHING.

Gauge, 4 ft., 8½ in. or wider.   Fuel, Bituminous Coal.

General Design shown by Plate XIV.

	Cylinders. Diameter and Stroke. inches.	Dia'eter of Driving Wheels. inches.	Wheel Base.		Weight, in running order. POUNDS.		Separate Tender.
			Of Driving Wheels.	Total.	On Driving Wheels. lbs.	Total. lbs.	Capacity of Tank. Gals.
1	14 × 22	50	10 ft. 3 in.	10 ft. 3 in.	64000	64000	1500
2	15 × 22	50	10 ft. 3 in.	10 ft. 3 in.	66500	66500	1500
3	16 × 20	50	10 ft. 3 in.	10 ft. 3 in.	67000	67000	1500
4	16 × 22	50	10 ft. 3 in.	10 ft. 3 in.	69000	69000	1800
5	16 × 24	50	10 ft. 7 in.	10 ft. 7 in.	73000	73000	1800

Load in tons of 2000 pounds in addition to Engine and Tender, at 10 miles an hour, on a grade per mile of

	On a Level.	10 ft.	20 ft.	40 ft.	60 ft.	80 ft.	100 ft.	125 ft.	150 ft.
1	2375	1504	1085	688	497	383	311	248	205
2	2450	1551	1120	710	513	397	322	256	210
3	2486	1575	1137	721	521	403	326	260	213
4	2559	1620	1169	741	535	413	335	266	218
5	2709	1716	1238	785	567	439	355	283	232

Six Wheel Locomotives

FOR SWITCHING.

Gauge, 4 ft., 8½ in. or wider.   Fuel, Bituminous Coal.

General Design shown by Plate XIV.

	Cylinders. Diameter and Stroke. inches.	Dia'eter of Driving Wheels. inches.	Wheel Base. Of Driving Wheels.	Wheel Base. Total.	Weight, in running order. POUNDS. On Driving Wheels. lbs.	Weight, in running order. POUNDS.	Weight, in running order. POUNDS. Total. lbs.	Separate Tender. Capacity of Tank. Gals.
1	17×22	50	10 ft. 7 in.	10 ft. 7 in.	74000		74000	1800
2	17×24	50	10 ft. 7 in.	10 ft. 7 in.	79000		79000	1800
3	18×22	50	10 ft. 7 in.	10 ft. 7 in.	80000		80000	2000
4	18×24	50	10 ft. 7 in.	10 ft. 7 in.	85000		85000	2000
5	19×22	50	10 ft. 7 in.	10 ft. 7 in.	86500		86500	2300

	Load in tons of 2000 pounds in addition to Engine and Tender, at 10 miles an hour, on a grade per mile of							
	On a Level.	10 ft.	20 ft.	40 ft.	60 ft.	80 ft.	100 ft.	125 ft.	150 ft.
1	2747	1740	1256	796	575	445	361	287	236
2	2933	1858	1341	851	615	476	386	307	253
3	2969	1880	1357	861	622	481	389	310	254
4	3155	1999	1443	915	661	512	415	330	271
5	3189	2019	1456	923	666	515	416	331	271

PLATE XV.

Four Wheel Tank Locomotives

FOR SWITCHING.

Gauge, 4 ft., 8½ in. or wider.   Fuel, Bituminous Coal.

General Design shown by Plate XV.

	Cylinders. Diameter and Stroke. inches.	Dia'eter of Driving Wheels. inches.	Wheel Base. Of Driving Wheels.	Total.	Weight, in running order. POUNDS. On Driving Wheels. lbs.		Total. lbs.	Tank on Engine. Capacity of Tank. Gals.
1	10½ × 18	34	6 ft. 6 in.	6 ft. 6 in.	43000		43000	500
2	11 × 18	46	6 ft. 6 in.	6 ft. 6 in.	46000		46000	500
3	14 × 22	46	7 ft. 6 in.	7 ft. 6 in.	54000		54000	550
4								
5								

	On a Level.	10 ft.	20 ft.	40 ft.	60 ft	80 ft.	100 ft.	125 ft.	150 ft.
	Load in tons of 2000 pounds in addition to Engine and Tender, at 10 miles an hour, on a grade per mile of								
1	1606	1021	740	473	345	269	220	177	148
2	1719	1093	793	507	369	289	236	190	158
3	2018	1283	930	595	434	339	277	223	186
4									
5									

Four Wheel Tank Locomotives

FOR SWITCHING.

Gauge. 4 ft.. 8½ in. or wider.   Fuel, Bituminous Coal.

General Design shown by Plate XV.

	Cylinders. Diameter and Stroke. inches.	Dia'eter of Driving Wheels. inches.	Wheel Base.		Weight, in running order. POUNDS.			Tank on Engine.
			Of Driving Wheels.	Total.	On Driving Wheels. lbs.		Total. lbs.	Capacity of Tank. Gals.
1	11 × 18	50	6 ft. 6 in.	6 ft. 6 in.	46800		46800	500
2	14 × 22	50	7 ft. 6 in.	7 ft. 6 in.	54800		54800	550
3	15 × 22	50	7 ft. 6 in.	7 ft. 6 in.	58500		58500	600
4	16 × 22	50	7 ft. 6 in.	7 ft. 6 in.	62000		62000	650
5								

	Load in tons of 2000 pounds in addition to Engine and Tender, at 10 miles an hour, on a grade per mile of								
	On a Level.	10 ft.	20 ft.	40 ft.	60 ft.	80 ft.	100 ft.	125 ft.	150 ft.
1	1756	1116	809	517	377	294	241	194	162
2	2055	1306	929	606	441	345	282	227	189
3	2204	1402	1016	650	473	370	302	244	203
4	2317	1473	1069	683	498	389	318	256	213
5									

PLATE XVI.

Six Wheel Tank Locomotives

FOR SWITCHING.

Gauge, 4 ft., 8½ in. or wider.   Fuel, Bituminous Coal.

General Design shown by Plate XVI.

	Cylinders. Diameter and Stroke. inches.	Dia'eter of Driving Wheels. inches.	Wheel Base.		Weight, in running order. POUNDS.			Tank on Engine.
			Of Driving Wheels.	Total.	On Driving Wheels. lbs.		Total. lbs.	Capacity of Tank. Gals.
1	11 × 18	46	10 ft.	10 ft.	50000		50000	500
2	14 × 22	46	10 ft.	10 ft.	70000		70000	500
3	15 × 22	46	10 ft.	10 ft.	73000		73000	600
4	16 × 22	46	10 ft.	10 ft.	76000		76000	650
5								

Load in tons of 2000 pounds in addition to Engine and Tender, at 10 miles an hour, on a grade per mile of

	On a Level.	10 ft.	20 ft.	40 ft.	60 ft.	80 ft.	100 ft.	125 ft.	150 ft.
1	1868	1188	863	551	402	316	257	207	172
2	2616	1664	1206	770	563	439	359	290	242
3	2728	1734	1257	804	586	458	374	302	252
4	2840	1806	1309	837	611	477	390	314	262
5									

Six Wheel Tank Locomotives

FOR SWITCHING.

Gauge, 4 ft., 8 ½ in. or wider.   Fuel, Bituminous Coal.

General Design shown by Plate XVI.

	Cylinders. Diameter and Stroke. inches.	Dia'eter of Driving Wheels. inches.	Wheel Base. Of Driving Wheels.	Total.	Weight, in running order. POUNDS. On Driving Wheels. lbs.		Total. lbs.	Tank on Engine. Capacity of Tank. Gals.
1	11 × 18	50	10 ft. 7 in.	10 ft. 7 in.	52000		52000	500
2	14 × 22	50	10 ft. 7 in.	10 ft. 7 in.	71600		71600	500
3	15 × 22	50	10 ft. 7 in.	10 ft. 7 in.	74600		74600	600
4	16 × 22	50	10 ft. 7 in.	10 ft. 7 in.	78000		78000	650
5								

	Load in tons of 2000 pounds in addition to Engine and Tender, at 10 miles an hour, on a grade per mile of								
	On a Level.	10 ft.	20 ft.	40 ft.	60 ft.	80 ft.	100 ft.	125 ft.	150 ft.
1	1943	1236	896	573	418	326	267	215	179
2	2691	1711	1240	793	578	452	369	298	248
3	2802	1782	1291	826	602	470	384	310	258
4	2915	1854	1344	860	626	489	400	323	269
5									

PLATE XVII.

Eight Wheel Double-Ender

TANK LOCOMOTIVE.

Gauge, 4 ft., 8½ in. or **wider**. Fuel, Bituminous Coal.

General Design shown by Plate XVII.

	Cylinders. Diameter and Stroke. inches.	Dia'eter of Driving Wheels. inches.	Wheel Base.		Weight, in running order. POUNDS.			Tank on Engine.
			Of Driving Wheels.	Total.	On Driving Wheels. lbs.	On Front and Rear Truck lbs.	Total. lbs.	Capacity of Tank. Gals.
1	13½ × 22	46	6 ft. 6 in.	20 ft. 9 in.	42200	{ 12000 9800	64000	700
2	15 × 22	46	6 ft. 8 in.	21 ft.	43500	{ 14000 10500	68000	700
3	16 × 22	46	7 ft.	21 ft. 6 in.	46500	{ 14500 11000	72000	750
4								
5								

Load in tons of 2000 pounds in addition to Engine and Tender, at 20 miles an hour, on a grade per mile of

	On a Level.	10 ft.	20 ft.	40 ft.	60 ft.	80 ft.	100 ft.	125 ft.	150 ft.
1	1239	847	636	419	308	241	197	157	130
2	1276	872	654	432	317	248	202	161	133
3	1364	933	700	461	339	265	215	173	142
4									
5									

Eight Wheel Double-Ender

TANK LOCOMOTIVE.

Gauge, 4 ft., 8½ in. or wider.   Fuel, Bituminous Coal.

General Design shown by Plate XVII.

	Cylinders. Diameter and Stroke. inches.	Dia'eter of Driving Wheels. inches.	Wheel Base.		Weight, in running order. POUNDS.			Tank on Engine.
			Of Driving Wheels.	Total.	On Driving Wheels. lbs.	On Front and Rear Truck lbs.	Total. lbs.	Capacity of Tank. Gals.
1	13½ × 22	50	6 ft. 6 in.	20 ft. 9 in.	43200	{ 12000 9800	65000	700
2	15 × 22	50	6 ft. 8 in.	20 ft.	44500	{ 14000 10500	69000	700
3	16 × 22	50	7 ft.	21 ft. 6 in.	47000	{ 14800 11200	73000	750
4								
5								

	Load in tons of 2000 pounds in addition to Engine and Tender, at 20 miles an hour, on a grade per mile of								
	On a Level.	10 ft.	20 ft.	40 ft.	60 ft.	80 ft.	100 ft.	125 ft.	150 ft.
1	1268	867	651	429	315	247	201	161	133
2	1305	892	669	440	324	253	206	165	136
3	1381	942	707	465	342	267	218	174	143
4									
5									

Eight Wheel Double-Ender

TANK LOCOMOTIVE.

Gauge, 4 ft., 8½ in. or wider.   Fuel, Bituminous Coal.

General Design shown by Plate XVII.

	Cylinders. Diameter and Stroke. inches.	Dia'eter of Driving Wheels. inches.	Wheel Base.		Weight, in running order. POUNDS.			Tank on Engine.
			Of Driving Wheels.	Total.	On Driving Wheels. lbs.	On Front and Rear Truck lbs.	Total. lbs.	Capacity of Tank. Gals.
1	15 × 22	56	6 ft. 9 in.	21 ft. 6 in.	45500	{ 14000 { 10500	70000	700
2	16 × 22	56	7 ft.	21 ft. 6 in.	48000	{ 14800 { 11200	74000	750
3								
4								
5								

Load in tons of 2000 pounds in addition to Engine and Tender, at 20 miles an hour, on a grade per mile of

	On a Level.	10 ft.	20 ft.	40 ft.	60 ft.	80 ft.	100 ft.	125 ft.	150 ft.
1	1335	915	685	451	332	260	212	169	140
2	1408	963	722	476	350	274	223	178	147
3									
4									
5									

PLATE XVIII.

Ten Wheel Double-Ender Locomotives

WITH TANK OVER REAR TRUCK.

Gauge, 4 ft., 8½ in. or wider.   Fuel, Bituminous Coal.

General Design shown by Plate XVIII.

	Cylinders. Diameter and Stroke. inches.	Dia'eter of Driving Wheels. inches.	Wheel Base.		Weight, in running order. POUNDS.			Tank on Engine.
			Of Driving Wheels.	Total.	On Driving Wheels. lbs.	On Front and Rear Truck lbs.	Total. lbs.	Capacity of Tank. Gals.
1	15 × 22	50	7 ft.	29 ft. 8 in.	60000	{ 10000 { 25000	95000	1000
2	16 × 22	50	7 ft.	29 ft. 8 in.	62000	{ 10500 { 25000	97500	1000
3								
4								
5								

Load in tons of 2000 pounds in addition to Engine and Tender, at 20 miles an hour, on a grade per mile of

	On a Level.	10 ft.	20 ft.	40 ft.	60 ft.	80 ft.	100 ft.	125 ft.	150 ft.
1	1759	1202	901	593	436	341	277	221	182
2	1818	1242	932	613	451	352	287	229	189
3									
4									
5									

Ten Wheel Double-Ender Locomotives

WITH TANK OVER REAR TRUCK.

Gauge, 4 ft., 8½ in. or wider.   Fuel, Bituminous Coal.

General Design shown by Plate XVIII.

	Cylinders. Diameter and Stroke. inches.	Dia'eter of Driving Wheels. inches.	Wheel Base.		Weight, in running order. POUNDS.			Tank on Engine.
			Of Driving Wheels.	Total.	On Driving Wheels. lbs.	On Front and Rear Truck lbs.	Total. lbs.	Capacity of Tank. Gals.
1	15 × 22	56	7 ft.	29 ft. 8 in.	61000	{ 10000 { 25000	96000	1000
2	16 × 22	56	7 ft.	29 ft. 8 in.	63000	{ 10500 { 25000	98500	1000
3								
4								
5								

	On a Level.	10 ft.	20 ft.	40 ft.	60 ft.	80 ft.	100 ft.	125 ft.	150 ft.
	Load in tons of 2000 pounds in addition to Engine and Tender, at 20 miles an hour, on a grade per mile of								
1	1789	1222	917	604	444	347	283	226	186
2	1847	1262	947	623	458	358	292	233	192
3									
4									
5									

PLATE XIX.

Ten Wheel Double-Ender Tank Locomotive

WITH SIX DRIVERS.

Gauge, 4 ft., 8 ½ in. or wider. Fuel, Bituminous Coal.

General Design shown by Plate XIX.

	Cylinders. Diameter and Stroke. inches.	Dia'eter of Driving Wheels. inches.	Wheel Base.		Weight, in running order. POUNDS.			Tank on Engine.
			Of Driving Wheels.	Total.	On Driving Wheels. lbs.	On Front and Rear Truck lbs.	Total. lbs.	Capacity of Tank. Gals.
1	15 × 20	42	12 ft.	24 ft.	67400	{ 9000 7500	83900	1400
2	15 × 20	49	12 ft. 6 in.	25 ft.	68000	{ 9300 7700	85000	1600
3								
4								
5								

	Load in tons of 2000 pounds in addition to Engine and Tender, at 20 miles an hour, on a grade per mile of								
	On a Level.	10 ft.	20 ft.	40 ft.	60 ft.	80 ft.	100 ft.	125 ft.	150 ft.
1	1988	1362	1024	678	502	394	323	261	217
2	2005	1373	1032	684	505	397	326	262	218
3									
4									
5									

PLATE XX.

Eight Wheel Forney Engine

WITH TANK OVER TRUCK.

Gauge, 4 ft., 8½ in. or wider.   Fuel, Anthracite Coal.

General Design shown by Plate XX.

	Cylinders. Diameter and Stroke. inches.	Dia'eter of Driving Wheels. inches.	Wheel Base.		Weight, in running order. POUNDS			Tank on Engine.
			Of Driving Wheels.	Total.	On Driving Wheels. lbs.	On Truck. lbs.	Total. lbs.	Capacity of Tank. Gals.
1	11 × 16	42	5 ft.	16 ft. 1 in.	29000	14000	43000	500
2	12 × 18	42	5 ft. 3 in.	16 ft. 7 in.	34000	18000	52000	650
3								
4								
5								

	Load in tons of 2000 pounds in addition to Engine and Tender, at 20 miles an hour, on a grade per mile of							
	On a Level.	10 ft.	20 ft.	40 ft.	60 ft.	80 ft.	100 ft.	125 ft.	150 ft.
1	851	582	437	288	217	166	135	108	89
2	998	682	512	337	248	194	158	127	104
3									
4									
5									

CHAPTER X.

PLATES AND TABLES OF DIMENSIONS AND CAPACITY OF NARROW GAUGE LOCOMOTIVES.

THE following are some of the styles of locomotives adapted to gauges of less than 4 feet 8½ inches:

In the construction of Narrow Guage Engines here shown, and more especially in the illustration of parts of Locomotives, it will be seen that to secure sufficient water space, steam room, and firebox room, special designs were made, which make these engines as efficient in service as those of wider gauge.

PLATE XXI.

Eight Wheel Standard Locomotives

FOR PASSENGERS OR FREIGHT.

Narrow Gauge Track.   Fuel, Bituminous Coal.

General Design shown by Plate XXI.

	Cylinders. Diameter and Stroke. inches.	Dia'eter of Driving Wheels. inches.	Wheel Base.		Weight, in running order. POUNDS			Separate Tender.
			Of Driving Wheels.	Total.	On Driving Wheels. lbs.	On Truck. lbs.	Total. lbs.	Capacity of Tank. Gals.
1	12 × 16	41	7 ft. 4 in.	19 ft.	26000	15000	41000	1200
2	13 × 20	55	7 ft. 9 in.	20 ft. 5 in.	29600	18200	47800	1300
3								
4								
5								

	Load in tons of 2000 pounds in addition to Engine and Tender, at 20 miles an hour, on a grade per mile of								
	On a Level.	10 ft.	20 ft.	40 ft.	60 ft.	80 ft.	100 ft.	125 ft.	150 ft.
1	748	507	376	243	175	134	106	81	65
2	851	577	428	278	199	152	121	93	74
3									
4									
5									

PLATE XXII.

Mogul Locomotives

FOR FREIGHT.

Narrow Gauge Track.   Fuel, Bituminous Coal.

General Design shown by Plate XXII.

	Cylinders. Diameter and Stroke. inches.	Dia'eter of Driving Wheels. inches.	Wheel Base.		Weight, in running order. POUNDS.			Separate Tender.
			Of Driving Wheels.	Total.	On Driving Wheels. lbs.	On Truck. lbs.	Total. lbs.	Capacity of Tank. Gals.
1	14 × 18	37	12 ft. 4 in.	18 ft. 4 in.	43600	8700	52300	1400
2	15 × 18	37	12 ft. 10 in.	18 ft. 10 in.	45500	9000	54500	1500
3								
4								
5								

	Load in tons of 2000 pounds in addition to Engine and Tender, at 20 miles an hour, on a grade per mile of								
	On a Level.	10 ft.	20 ft.	40 ft.	60 ft.	80 ft.	100 ft.	125 ft.	150 ft.
1	1271	866	648	425	310	240	194	154	125
2	1326	904	676	442	323	251	203	160	131
3									
4									
5									

Mogul Locomotives

FOR FREIGHT.

Narrow Gauge Track.   Fuel, Bituminous Coal.

General Design shown by Plate XXII.

	Cylinders. Diameter and Stroke. inches.	Dia'eter of Driving Wheels. inches.	Wheel Base.		Weight, in running order. POUNDS.			Separate Tender.
			Of Driving Wheels.	Total.	On Driving Wheels. lbs.	On Truck. lbs.	Total. lbs.	Capacity of Tank. Gals.
1	14 × 18	41	13 ft.	19 ft. 3 in.	44500	8700	53200	1400
2	15 × 18	41	13 ft. 4 in.	19 ft. 7 in.	47000	9000	56000	1500
3								
4								
5								

	Load in tons of 2000 pounds in addition to Engine and Tender, at 20 miles an hour, on a grade per mile of								
	On a Level.	10 ft.	20 ft.	40 ft.	60 ft.	80 ft.	100 ft.	125 ft.	150 ft.
1	1297	884	661	432	316	245	193	157	128
2	1373	936	699	457	334	259	210	166	136
3									
4									
5									

PLATE XXIII.

Mogul Locomotives

FOR FREIGHT.

Narrow Gauge Track.   Fuel, Bituminous Coal.

General Design shown by Plate XXIII.

	Cylinders. Diameter and Stroke. inches.	Dia'eter of Driving Wheels. inches.	Wheel Base.		Weight, in running order. Pounds.			Separate Tender.
			Of Driving Wheels.	Total.	On Driving Wheels. lbs.	On Front Truck On Each Tender Trk lbs.	Total. lbs.	Capacity of Tank. Gals.
1	13 × 18	30	7 ft. 6 in.	33 ft. 3 in.	38000	{ 9500 { 12000	75000	1000
2	13 × 18	37	7 ft. 6 in.	33 ft. 9 in.	39500	{ 9500 { 12000	77000	1000
3								
4								
5								

	On a Level.	10 ft.	20 ft.	40 ft.	60 ft.	80 ft.	100 ft.	125 ft.	150 ft.
	Load in tons of 2000 pounds in addition to Engine and Tender, at 20 miles an hour, on a grade per mile of								
1	1100	748	557	362	262	202	162	126	102
2	1144	778	580	377	273	211	170	132	107
3									
4									
5									

In this style of Engine the rear end of the Engine is connected to and supported by the front end of the Tender.

The forward tender truck is provided with swing motion.

These Engines are specially adapted to sharp curves.

PLATE XXIV.

Four Wheel Tank Locomotive

FOR SWITCHING.

Narrow Gauge Track.   Fuel, Bituminous Coal.

General Design shown by Plate XXIV.

	Cylinders. Diameter and Stroke. inches.	Dia'eter of Driving Wheels. inches.	Wheel Base. Of Driving Wheels.	Wheel Base. Total.	Weight, in running order. POUNDS. On Driving Wheels. lbs.		Total. lbs.	Tank on Engine. Capacity of Tank. Gals.
1	8 × 12	26	5 ft.	5 ft.	18000		18000	175
2	8 × 12	30	5 ft.	5 ft.	18500		18500	175
3								
4								
5								

Load in tons of 2000 pounds in addition to Engine and Tender, at 10 miles an hour, on a grade per mile of

	On a Level.	10 ft.	20 ft.	40 ft.	60 ft.	80 ft.	100 ft.	125 ft.	150 ft.
1	673	428	310	198	144	113	92	74	62
2	692	440	319	204	149	116	95	77	64
3									
4									
5									

Four Wheel Tank Locomotive

FOR SWITCHING.

Narrow Gauge Track.   Fuel, Bituminous Coal.

General Design shown by Plate XXIV.

	Cylinders. Diameter and Stroke. inches.	Dia'eter of Driving Wheels. inches.	Wheel Base.		Weight, in running order. POUNDS.			Tank on Engine.
			Of Driving Wheels.	Total.	On Driving Wheels. lbs.		Total. lbs.	Capacity of Tank. Gals.
1	9 × 16	30	5 ft. 3 in.	5 ft. 3 in.	28000		28000	275
2	9 × 16	37	5 ft. 3 in.	5 ft. 3 in.	29000		29000	275
3								
4								
5								

	Load in tons of 2000 pounds in addition to Engine and Tender, at 10 miles an hour, on a grade per mile of								
	On a Level.	10 ft.	20 ft.	40 ft.	60 ft.	80 ft.	100 ft.	125 ft.	150 ft.
1	1046	666	482	308	225	176	144	116	96
2	1083	689	499	319	232	181	148	120	99
3									
4									
5									

PLATE XXV.

Mogul Tank Locomotive

FOR FREIGHT.

Narrow Gauge Track.   Fuel, Bituminous Coal.

General Design shown by Plate XXV

	Cylinders. Diameter and Stroke. inches.	Dia'eter of Driving Wheels. inches.	Wheel Base. Of Driving Wheels.	Total.	Weight, in running order. POUNDS. On Driving Wheels. lbs.	On Truck. lbs.	Total. lbs.	Tank on Engine. Capacity of Tank. Gals.
1	13 × 18	30	9 ft.	13 ft. 10 in.	41000	8000	49000	740
2								
3								
4								
5								

Load in tons of 2000 pounds in addition to Engine and Tender, at 20 miles an hour, on a grade per mile of

	On a Level.	10 ft.	20 ft.	40 ft.	60 ft.	80 ft.	100 ft.	125 ft.	150 ft.
1	1209	829	624	413	306	241	197	159	132
2									
3									
4									
5									

PLATE XXVI.

Eight Wheel Double-Ender Locomotive

FOR FREIGHT OR PASSENGERS

Narrow Gauge Track.   Fuel, Bituminous Coal.

General Design shown by Plate XXVI.

	Cylinders. Diameter and Stroke. inches.	Dia'eter of Driving Wheels. inches.	Wheel Base. Of Driving Wheels.	Wheel Base. Total.	Weight, in running order. POUNDS. On Driving Wheels. lbs.	Weight, in running order. POUNDS. On Front and Rear Truck lbs.	Weight, in running order. POUNDS. Total. lbs.	Separate Tender. Capacity of Tank. Gals.
1	12 × 20	44	6 ft.	22 ft. 1 in.	31000	{ 11000 9000	51000	1500
2	12 × 20	49	6 ft.	22 ft. 1 in.	32000	{ 11000 9000	52000	1500
3								
4								
5								

	Load in tons of 2000 pounds in addition to Engine and Tender, at 20 miles an hour, on a grade per mile of								
	On a Level.	10 ft.	20 ft.	40 ft.	60 ft.	80 ft.	100 ft.	125 ft.	150 ft.
1	908	620	465	305	224	175	142	113	93
2	938	641	480	316	232	181	147	118	97
3									
4									
5									

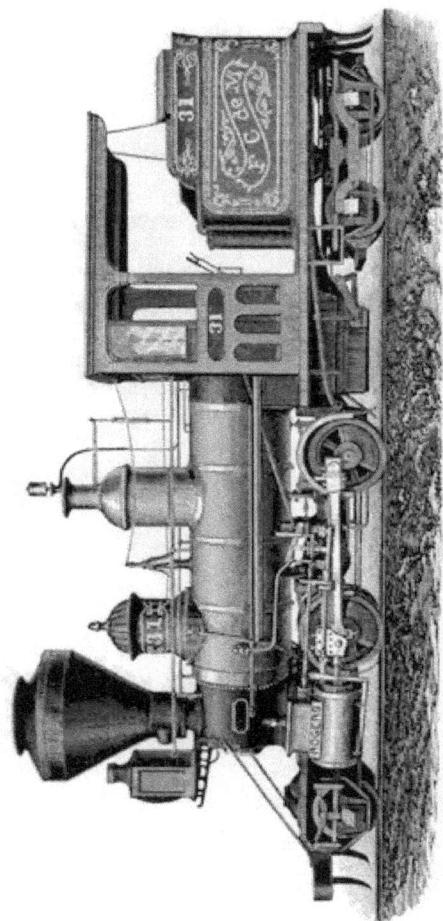

PLATE XXVII.

Ten Wheel Double-Ender Locomotive

WITH TANK OVER REAR TRUCK.

Narrow Gauge Track. Fuel, Bituminous Coal.

General Design shown by Plate XXVII.

	Cylinders. Diameter and Stroke. inches.	Dia'eter of Driving Wheels. inches.	Wheel Base. Of Driving Wheels.	Total.	Weight, in running order. POUNDS. On Driving Wheels. lbs.	On Front and Rear Truck lbs.	Total. lbs.	Tank on Engine. Capacity of Tank. Gals.
1	9 × 12	30	6 ft.	24 ft.	20000	{ 5000 14000	39000	600
2	9 × 12	36	6 ft.	26 ft.	21500	{ 5000 14000	40500	600
3								
4								
5								

Load in tons of 2000 pounds in addition to Engine and Tender, at 20 miles an hour, on a grade per mile of

	On a Level.	10 ft.	20 ft.	40 ft.	60 ft.	80 ft.	100 ft.	125 ft.	150 ft.
1	582	397	296	194	141	109	88	70	56
2	624	425	318	209	152	119	96	76	62
3									
4									
5									

PLATE XXVIII.

Fourteen Wheel Double-Ender Locomotive

WITH SIX DRIVERS AND TANK OVER REAR TRUCK.

Narrow Gauge Track.   Fuel, Bituminous Coal.

General Design shown by Plate XXVIII.

	Cylinders. Diameter and Stroke. inches.	Diameter of Driving Wheels. inches.	Wheel Base. Of Driving Wheels.	Wheel Base. Total.	Weight, in running order. Pounds. On Driving Wheels. lbs.	Weight, in running order. Pounds. On Front and Rear Truck. lbs.	Weight, in running order. Pounds. Total. lbs.	Tank on Engine. Capacity of Tank. Gals.
1	12 × 16	34	9 ft. 3 in.	31 ft. 1½ in.	39000	8000 25000	72000	1200
2	13½ × 16	34	9 ft. 3 in.	31 ft. 1½ in.	40000	8200 26000	74200	1300
3								
4								
5								

Load in tons of 2000 pounds in addition to Engine and Tender, at 20 miles an hour, on a grade per mile of

	On a Level.	10 ft.	20 ft.	40 ft.	60 ft.	80 ft.	100 ft.	125 ft.	150 ft.
1	1138	776	581	380	278	216	175	139	113
2	1197	790	595	390	285	222	179	142	116
3									
4									
5									

INDEX.